# Lecture Notes in Morphogenesis

**Series editor**

Alessandro Sarti, CAMS Center for Mathematics, CNRS-EHESS, Paris, France
e-mail: alessandro.sarti@ehess.fr

More information about this series at http://www.springer.com/series/11247

Alessandro Sarti · Federico Montanari
Francesco Galofaro
Editors

# Morphogenesis and Individuation

Springer

*Editors*
Alessandro Sarti
CAMS Center for Mathematics
CNRS-EHESS
Paris
France

Francesco Galofaro
CUBE, Università di Bologna
Bologna
Italy

Federico Montanari
Università di Urbino
Urbino
Italy

ISSN 2195-1934 ISSN 2195-1942 (electronic)
ISBN 978-3-319-05100-0 ISBN 978-3-319-05101-7 (eBook)
DOI 10.1007/978-3-319-05101-7

Library of Congress Control Number: 2014949297

Springer Cham Heidelberg New York Dordrecht London



Printed on acid-free paper

Springer is part of Springer Science+Business Media (www.springer.com)

# Foreword

This volume, 'Morphogenesis and Individuation', aims to build on the categories and conceptual tools used in the morphogenetic approaches through a discussion of Gilbert Simondon's work. Simondon's philosophical perspectives on science were further developed in René Thom's mature analysis. Key-concepts as information, metastability and individuation opened new issues for discussion. In particular, even though Thom's concept of morphodynamics provided a topologic categorisation of the spatio-temporal configurations, it left several problems open, such as the constitution of singularities starting from a multiplicity of elements, as well as the genesis and evolution of the space of control, that in Thom's approach is a priori given. The category of *individuation* plays an important role in Thom's framework, explaining the relationships between *saliencies,* i.e. perceived Gestalts, and what he calls a *prégnance,* a biological or physical signification which an organism, for innate or conditioned reasons, may ascribe to such salient forms—cf. Bundgaard and Stjernfelt (2010). Nevertheless, he admits that the issue of the individuation of a general *prégnance* is obscure, leaving only conjectural hypotheses—cf. Thom (1991).

For this reason, a closer look at the concept of individuation in Simondon's work provides the opportunity for new research and developments in different areas, and could provide a unitary framework for disciplines interested in perception, vision, language, anthropology, semiotics and cultural studies.

Thom (1968) applied morphogenetic models to the morphodynamic development of syntax. The laws of form generation and stability lead to simple archetypes that link language with our experience of the environment, semiotics, biology, culture and nature. These models describe meaning through different manners of capturing salient objects by pregnant 'capture devices'. This concept of morphology was further developed by Wildgen (1982).

The issue of the individuation of singular *prégnances* from a general one was posed by Thom in terms of the generation of the four base-relations between actants starting from the hysteresis cycle—cf. Thom (1989, 1991). The relation between the individuation of a general prégnance and the archetypal syntax recalls Greimas' distinction between fundamental and narrative syntax—cf. Greimas and

Courtés (1979)—suggesting a mathematical model that generates both of them. The boundaries between Semiotics and Morphogenetic approaches has also been analysed by Petitot (2003).

Thom's concepts of individuation and metastability are related to Simondon's work. However, compared to Simondon's concept of individuation, Thom's models seem too deterministic. A study of the organisational and perceptual faculties of living beings, of their topologic and functional plasticity, and their complex behaviour should avoid the mechanistic, deterministic and reductionist perspectives that have been popular in the last 50 years. The morphogenetic perspective, instead, could establish a new fruitful research direction by reconsidering the Simondonian concept of individuation, and its relationship to indetermination and identity. This could lead to a non-deterministic approach to meaning construction, a new way to address the link between local and global processes, as well as describe phenomena in their diachronic variation.

Simondon's work on individuation (2005) indicates a way for forms to find an internal consistency and to become actualized. Simondon considers information processes as the acquisition of form, generated by the metastability of the system. Individuation is a movement of resonance and internal reconfiguration in which signals and messages connect heterogeneous pre-individual differences, integrating their resonances and thus generating global differences. Naturally, individuation is always an incomplete and partial process, a variable network of both pre-individual and singular features.

Simondon considered certain key-concepts the starting point for a possible axiomatisation of a general epistemology for the human sciences: form, information, potential, transduction—cf. 'introduction' in Simondon (1989). 'Form' is an active principle that operates on matter. Nevertheless, in Simondon's perspective matter is not entirely passive. Matter is seen as a force field in which the *germ* of the form can propagate thanks to the metastable state of the matter. The matter is the locus of the potentials, and the boundary between form and matter is an amplifier signal. The propagation of the form is called 'transduction'. Thus, Simondon reforms Aristotle's old ileomorphic schema by introducing a morphodynamical perspective. Simondon's perspective is very different from Thom's. As Andrea Bardin underlines in his paper, the starting point of Simondon's model is not the catastrophic segmentation of an original continuum, but the propagation of singular discontinuities.

It is important to distinguish between the different possible metastable states, which are responsible for the possibility of a hierarchy of good forms, and the stable state, which coincides with death. In the stable state no more changes are possible in this system without the intervention of external energies. Using this definition Simondon associates systems with a certain degree of (meta)stability.

With this model in mind, Simondon conducts a philosophical shift. He considers 'individuation' a resolution starting from a metastable state. In other terms, he does not try to explain 'individuation' starting from the individual. He does not take the existence of well-delimited individuals for granted (each one with its 'haecceitas'), as we find in philosophy from the middle-ages. Thanks to the fact that we can distinguish different degrees of stability and chaos, we can investigate the individual through the dynamics of individuation, starting from a pre-individual reality. The individual preserves its original bounds within the pre-individual and can never be considered an isolated monad. On the contrary, the collective, trans-individual dimension is already inscribed onto the individual.

This is why it is impossible to think about the individuation process using classical logic. Principles such as the Law of identity and of excluded-middle, already imply well individuated objects—Simondon finds an alternative in Quantum theory.

Simondon's work had important resonance in different fields due to his attempt to reformulate ontogenesis, not in mere individual terms, but as the becoming of being. In epistemological terms, this forced human sciences to integrate every morphology with an *energetics*. Even if many aspects of his thoughts were never fully understood, this fascinating perspective deeply influenced René Thom's and Gilles Deleuze's work.

Deleuze (1991:247, 318 n. 25) was interested in Simondon's definition of individuation, which presupposes a prior metastable state that is characterised by the distance between heterogeneous orders, as two or more structural series. Deleuze interprets their internal resonances as the constitution of systems. In this way, Simondon's model of individuation provides us with clues regarding the constitution of the plane of immanence, the space in which meaning is generated. 'Resonance' is the way in which heterogeneous concepts can be integrated into a non-fragmented whole (Deleuze and Guattari 1994:35). According to the deleuzian hypothesis, meaning extends from an 'immersive' situation (the 'immanence plane') in which processes and intensities work and produce sense relations without 'a priori forms'. How can we describe the immanence plane? A model that describes the constitution of the plane of immanence could be an important instrument, also directly related to the possibility of describing the generation of meaning and its structural articulation. In fact, describing the immanence plane is a risky operation. 'When the subject or the object falling outside the plane of immanence is taken as a universal subject or as any object *to which* immanence is attributed, the transcendental is entirely denatured, for it then simply redoubles the empirical (as with Kant), and immanence is distorted, for it then finds itself enclosed in the transcendent'—Deleuze (2001:27).

The essays have been subdivided in three thematic nuclei: (1) *Rethinking Individuation and Morphogenesis*; (2) *Morphologies, Culture and Spaces* and (3) *Immanence in Semiotics*. We start from the relationship between Thom's morphogenetic models and Simondon's notion of individuation. In particular, simondonian notions are often utilised in Thom's later works. For this reason, the first part of the volume focuses on Simondon's philosophy, hoping to cast new light in the direction of resonant and vibrational morphogenetic processes.

According to Andrea Bardin, Simondon's model of individuation is an alternative both to deterministic and in-deterministic perspectives in the philosophy of science. In fact, Simondon refuted Cartesian dualism, which indicates that the institution of the sciences depends on an ontological difference between the human being and nature. This implies free will on one side, and a sort of neo-mechanism on the other. Simondon also rejected the position that states that meaning is present in discourse and therefore in nature, such as in phenomenological philosophy. According to Simondon, individuation explains how being-as-a-subject and being-as-an-object come from the same primitive reality. The conditions for knowledge possibility and the causes of individuated being's existence are the same; this confirms the universality of knowledge.

The indetermination of the individuation process is the focus of Giovanni Carrozzini's article, who deeply analyses the constitution of perceptive forms in Simondon's epistemology.

According to Simondon, during the operation of perception, perceptive forms are invented particularly due to elements that cannot be reduced to a principle of order and simplicity. From this perspective, perceptive forms are not the simple product of the application of innate and determined schemes. They are, on the contrary, an inventive result that includes 'undetermined' elements. This chapter explains the reasons for Simondonian criticisms of associationism and Gestalt theory, and explains his original proposal to modify the concept of pregnant perceptive forms, by taking the singular-contextual elements of the perceptual experience into account.

Alessandro Sarti and David Piotrowski's chapter continues the discourse of the constitution of perceptual units, by proposing a model for the internal processes which takes place during the operation of individuation. They outline the common perspectives in the work of Simondon, Deleuze-Guattari and Bateson, framing the individuation principle inside a relational epistemology. This allows us to consider the pre-individual as a heterogeneous continuously changing relational field, which is functionally supported by harmonic processes that individuate its consistent forms. With this framework in mind, the authors intend to highlight the functional/operatorial level of the individuation process in terms of its dual articulation of: (a) the 'definition' of contextual-immanent relational graphs and (b) the estimation of their reduced spectral structure by means of non-linear harmonic analysis. This research indicates that the individuation process concerns not only forms, but also spaces (planes of consistencies) in general. In the end, the relationship between individuated spaces and the space of control from the Thomian morphodynamical tradition is outlined, observing that the individuated space provides the axes for the space of control. Suitable (arbitrary) potentials complete the construction of the space of control, allowing for semiotic oppositions and categorization. In this manner the space of control is not an a priori given, but is constantly formed through a morphogenetic process.

Claudia Mongini's chapter broadens the perspective of individuation from perception to perceptibility. The emergence of sensibility is the focus of this dense research. Sensibility can be understood as the perceptibility of objects, phenomena and processes, which have not yet emerged on the conscious level, and consequently do not possess a stable and recognisable form within the cognitive configuration. The aesthetic dimension unveils the composition of the immanent and contextual relational field that defines pre-individualility. Claudia Mongini utilizes

Simondon's theory that links the fields of technique and aesthetics through the concept of techno-aesthetics. In this manner, aesthetics express the way a tool adapts to its function. It is grasped in its operational condition towards action, and denotes the coming of a surprising event, instead of the predictable and repetitive state. Aesthetics becomes a bridge in a specific co-adaptation between conflicting experiences, acquiring both a functional and an operative character.

The Part II of the volume presents three research projects that demonstrate the great epistemological potential of the morphogenetic approach for different fields: anthropology; paleolinguistics; and cultural geography. In a thomian perspective, natural and cultural changes are never in conflict, and are always coupled without reducing one to the other. Individuation expresses their connection. In his article 'The cultural individuation of human language capacity and the morphogenesis of basic argument-schemata', Wolfgang Wildgen describes how basic human capacities such as language capacity, writing and cultural innovations can not be explained by neo-darwinian theories due to the short lapse of time in which they arise. Mophogenesis and individuation help explain the relationship between genotype and phenotype without rejecting Darwin's principles. For example, individuation can represent a bridge between biological and cultural innovation. Individuals with new features must survive in the environment of individuals who do not possess these changes in order to be replicated. Similarly, cultural innovation must be perceived as positive by other individuals in order to be imitated. Morphogenetic processes explain the unfolding of specific *prégnances*, i.e. in the case of language and symbolic development. This differentiation is reflected in the syntax; syntactic archetypes can be interpreted in terms of mental scenarios. These cognitive schemas explain a number of activities. For example, the catastrophe of capture can explain how a person controls a rabbit or a stone, opening interesting perspectives on the dynamics of cultural development in the neolithic period.

The article 'Through the Looking-Map: Mapping as a Milieu of Individuation' by Mario Neve shows how the simondonian notion of individuation is crucial in explaining the mapping processes. The author describes 'mapping' as a transduction: the content C is represented with a form F′ through a different form F (e.g. the euclidean geometry). The human animal is not opposed to the environment, because it is a part of it. Human individuation is never complete: the human animal is always in a metastable state. This feature, which distinguish the individuation of the living being from the individuation of technical objects, expresses the historical dynamic which allows the actualization of the world as a product of historical and social choices, as well as the crisis of a given actualization, which leads to a new virtualisation of the environment.

Maps are also a keyword in Ferraro's article 'On Growth and Form of Narrative Structures'. Ferraro discusses the narratological concept of narrativity as a linear transformation (e.g. from 'prisoner' to 'fugitive') by arguing that the transforming function takes place within a map of meaningful possibilities. We can consider 'narrativity' as the complex, unstable relationship between the route and the map, the process and the system. Therefore, a more complex conception of transformation is required, such as that which Lévi-Strauss borrowed from D'Arcy

Thompson's work. In light of the concept of transformations between system, the search for transformational groups implies a complex notion of culture; not only a collection of texts but a network of related textual fragments, which come with multiple identities, in reference to different systems.

The Part III of the volume reflects on the constitution of the immanence plan. In his chapter, Francesco Marsciani reconstructs the path that led the Danish linguist and semiotician L. Hjelmslev to reconstruct the science of language by only using concepts that are immanent to it; avoiding sociological, psychological or philosophical unnecessary premises. These concepts are organised by a metalanguage in a consistent plane, which represents the formal condition of possibility for meaning, independently from the substance that 'fits' the empty squares of this formal space. As a result of this operation, this space is constructed as transcendental. Nevertheless, there is no dichotomy between form and substance. Otherwise, form would have logical, transcendent laws, whose substance would only be realised without playing an active role. On the contrary, form is the result of the theoretical control of the transformation of the substance.

As Marsciani writes, the immanent criterion would ask the transcendental to not transcend phenomena, and the transcendental criterion would force immanence to take form and not submit to the substantive adventures of the phenomenon to which it belongs. Marsciani underlines how a relational epistemology, which takes the transcendental aspect of the immanent forms into account, does not need other foundations. It does not allow us to think about science and its objects as different things. We have already seen, with Bardin, how Simondon found a similar solution.

Building on Simondon's conceptions of individuation, metastability and indetermination, Galofaro's work presents meaning as a form *(Gestalt)* that can be individuated through a process of formation *(Bildung)* from a pre-individual, previously undetermined semantic universe. The author works with Thom's archetypal morphology in two directions. On one hand he uses quantum computation in order to represent indetermination. And on the other he imports the addressing function from Greimas' theory in order to represent the introduction and the circulation of semantic values into the semantic universe. This starts from a transcendent space that is part of the immanence plan, as both substance and modes are in immanence according to Spinoza. This model of *transduction* is a specific case. Quantum computation is currently seen as a discrete finite case of quantum field computation where the number of qubits is finite. This view coincides with Simondon's consideration of matter as a field. Nevertheless, this model indicates many possible implications as to the connections between the immanence plane and our experience of the *Lebenswelt.*

Federico Montanari, in his chapter, investigates the relationships between Deleuze's plane of immanence and Simondon's individuation. Montanari highlights the important role played by individuation in Deleuze's theory, as well as Spinoza's concept of expression and Hjelmslev's epistemology. As the author suggests, the political and philosophical consequences of these relationships constitute an innovative perspective for semiotic studies.

In conclusion, Simondon, Deleuze and Thom's latest research reformulate the morphogenetic perspective, introducing the conceptions of individuation, prégnance/salience and metastability. We attempt to clarify this framework, and redefine it in the light of a relational epistemology, as well as regarding contemporary concepts of functional plasticity.

## References

Berardi, F., Sarti, A.: Morphogenetic vibration. European Conference SCEPSI, San Marino (2011) http://www.slashdocs.com/vtvhn/morphogenetic-vibration.html.

Bundgaard, P.F., Stjernfelt, F.: René Thom semiotics and its sources. In: Wildgen, W., Brandt, P.A. (eds.) (2010)

Deleuze, G.: Difference and Repetition. Columbia University Press, New York (1991)

Deleuze, G.: Pure Immanence. Essays on A Life. Urzone Inc., New York (2001)

Deleuze, G., Guattari, F.: What is Philosophy. Columbia University Press, New York (1994)

Galofaro, F.: We have to change mind. Neural plausibility and the crisis of cognitive explanations. In: RIFL SFL pp. 101–115 (2011)

Greimas, Algirdas, J., Courtés, Joseph: Sémiotique: dictionnaire raisonné de la théorie du langage, Paris: Seuil, English trans. Semiotics and Language: An Analytical Dictionary. Indiana University Press, Indiana (1983) (1979)

Montanari, F.: Language, Culture, Vision. Some Ideas for a Critical Approach. In RIFL SFL pp 180–191 (2012)

Montanari, F.: Between trees, webs and mirrors. Dimensions of immanence and a critical poststructuralist proposals, in E/C, http://www.ec-aiss.it/index_d.php?recordID=656 (2012)

Petitot, J.: Morphogenesis of Meaning. Peter Lang, Bern (2003)

Sarti, A., Citti, G.: On the origin and nature of neurogeometry. In: La Nuova Critica (2011)

Simondon, G.: L'individuation psychique et collective à la lumière des notions de Forme, Information, Potentiel et Métastabilité. Aubier, Paris (1989)

Simondon, G.: L'individuation à la lumiere des notions de formes et d'information. Jérôme Millon, Grenoble (2005)

Thom, R.: Topologie et signification. In: L'Age de la science, 4. (1968) (Trans. in Thom (1983))

Thom, R.: Mathematical Models of Morphogenesis. Horwood, New York (1983)

Thom, R.: Cyclical structures in semiotics. In: Perron, Collins (eds.) Paris School Semiotics. John Benjamins Publishing Company, Amsterdam (1989)

Thom, R.: Morphologie du sémiotique, in Apologie du Logos. Hachette, Paris (1991)

Wildgen, W.: Catastrophe theoretic semantics. An Application and Elaboration of René Thom's Theory. Benjamins, Amsterdam (1982)

Wildgen, W., Brandt, P.A. (eds.), Semiosis and catastrophes. René Thom's Semiotic Heritage. Peter Lang, Bern (2010)

# Contents

**Part I Rethinking Individuation and Morphogenesis**

1 **On Substances and Causes Again: Simondon's Philosophy of Individuation and the Critique of the Metaphysical Roots of Determinism** . . . . . 3
Andrea Bardin

2 **How to Invent a Form: An Inquiry into Gilbert Simondon's Philosophy of Perception** . . . . . 33
Giovanni Carrozzini

3 **Individuation and Semiogenesis: An Interplay Between Geometric Harmonics and Structural Morphodynamics** . . . . . 49
Alessandro Sarti and David Piotrowski

4 **Morphogenesis Under Construction: Tracing the Process of Individuation Along Physico-Aesthetic Coordinates** . . . . . 75
Claudia Mongini

**Part II Morphologies, Culture, and Spaces**

5 **The Cultural Individuation of Human Language Capacity and the Morphogenesis of Basic Argument-Schemata** . . . . . 93
Wolfgang Wildgen

6 **Through the Looking-Map: Mapping as a Milieu of Individuation** . . . . . 111
Mario Neve

7 **On Growth and Form of Narrative Structures** . . . . . . . . . . . . . . . . 141
Guido Ferraro

**Part III Immanence in Semiotics**

8 **Semiotic Immanence and Formal Options** . . . . . . . . . . . . . . . . . . . 165
Francesco Marsciani

9 **Structural Syntax and Quantum Computation: A Simondonian Approach** . . . . . . . . . . . . . . . . . . . . . . . . . . . . . . . . . . . . . . . . 173
Francesco Galofaro

10 **Immanence/Imminence. Thinking About Immanence and Individuation** . . . . . . . . . . . . . . . . . . . . . . . . . . . . . . . . . . . 203
Federico Montanari

# Part I
# Rethinking Individuation and Morphogenesis

# Chapter 1
# On Substances and Causes Again: Simondon's Philosophy of Individuation and the Critique of the Metaphysical Roots of Determinism

**Andrea Bardin**

**Abstract** In his main work, *L'individuation à la lumière des notions de forme et d'information*, Gilbert Simondon displayed a theory of the discontinuous processes of individuation (or 'ontogenesis') from which structures emerge. Linking the concepts of singularity and historicity through the paradigmatic assumption of quantum physics, Simondon attacked both determinism and indeterminism by way of an original critique—neither empiricist nor idealistic—of the concepts of substance and cause.

> Il faut arriver à dissoudre cet énorme bloc du déterminisme métaphysique qui pèse sur la pensée scientifique
> Gaston Bachelard, *Le nouvel esprit scientifique*

## 1.1 Introduction

In his main work, *L'individuation à la lumière des notions de forme et d'information*,[1] Gilbert Simondon displayed a theory of the discontinuous processes of individuation (or 'ontogenesis') from which structures emerge. In Simondon's book one can only find sparse and rare references to topology: 'topology' is there the name Simondon gives to a new, complex mechanistic approach for the understanding and the explanation of any kind of process (physical, biological and

[1] Simondon completed the book in 1957, as his main PhD thesis. At the time two theses were required for completion of a PhD in French academia. Simondon's second dissertation was *Du mode d'existence des objets techniques*. While *Du Mode* was immediately published in 1958, thus making Simondon known as a philosopher of technology, *Individuation* underwent a quite complicated editorial process (see note 17). In what follows I will refer to Simondon's main work as simply *Individuation* and quote it as ILFI, according to common scholarly citation.

A. Bardin (✉)
Brunel University, London, UK
e-mail: bardin.andrea@gmail.com

 
A. Sarti et al. (eds.), *Morphogenesis and Individuation*,
Lecture Notes in Morphogenesis, DOI 10.1007/978-3-319-05101-7_1

psycho-social). It is a term he evokes in order to make a point against the deterministic form of mechanism which has shaped modern mechanicism since its beginnings. Although Simondon, during the 1980s, attended René Thom's seminars, and the latter dedicated a short article to the former after his death (Thom 1994), the absence of any direct reference to Thom's writings in Simondon's books shows no particular evidence of an actual historical link between the two thinkers. And, nevertheless, René Thom's brief essay *Halte au hasard, silence au bruit* triggered an interesting dispute on determinism in the 1980s, *La querelle du déterminisme* (Amsterdamski et al.1990),[2] which can be a valid step to understanding what Simondon was concerned with when he elaborated his philosophy of individuation. A brief *detour* through this dispute will provide a standpoint from which to appreciate how Simondon's theory of individuation contributes to a criticism of metaphysical assumptions which, unexpectedly, inhabit the most anti-metaphysical stances in what Althusser (1974) called the 'spontaneous philosophy of scientists'. Linking the concepts of singularity and historicity through the paradigmatic assumption of quantum physics, Simondon formulates a peculiar conception of transductive processes, which allows him to attack both determinism and indeterminism by way of an original critique—neither empiricist nor idealistic—of the concepts of substance and cause. In fact, the approach through which Simondon challenges morphogenetic processes in *Individuation* is effective both at the epistemological and at the historico-philosophical level. Taking his stand on his master Canguilhem's assumptions, Simondon's philosophy of individuation contributes both to dismantle the undisputed premises of the *querelle* on determinism themselves and to reveal the very metaphysical nature of modern mechanistic ontology.

## 1.2 Fascination with *Clinamen*

René Thom's article *Halte au hasard, silence au bruit* (1990a) harshly attacked what he polemically named the 'French popular epistemology', in which he included Edgar Morin, Jacques Monod and the 'couple' Prigogine-Stengers, i.e. those who 'sophistically' defended the notion of chance, captured as they were by some 'fascination with *clinamen*'.[3] On the contrary, according to Thom, a metaphysical 'decision' for determinism is the only one possible for an ethics capable of driving the asymptotic progress of science:

> As a philosopher, the scientist can leave the question open, but as a scientist, it is a question of principle for him [...] to adopt an optimistic perspective, postulating that nothing, in nature, is a priori unknowable. (Thom 1990a, p. 63)

---

[2] Besides Thom's essay and the subsequent debate published in the review *Le débat*, the book contains some further additions where the interlocutors revisited their arguments a few years later.

[3] According to Thom, chance is 'a void concept', a 'substitute of divine finality' (Thom 1990a, p. 75, 63).

From Thom's essay one understands that the metaphysical decision for ontological determinism is the proper founding value of an ethics of science, which opens and defines the horizon of scientific research, thus circumscribing all of its possible empirical limitations.

After 6 years Thom further clarified his view in another brief text entitled *Postface au debat sur le determinisme* (1990b). There he initially seemed to accept Amsterdamski's differentiation between 'global determinism' and 'local determinism' (Amsterdamski 1990), admitting that it is not possible to decide on a global determinism, since the object of scientific research is always limited and therefore the determinism one can derive from it is necessarily partial, local. These considerations brought him to progressively reformulate his initial claims and to eventually chose a kind of *epistemological determinism* in which a deterministic postulate still grounds the ethics of science without concerning the *ontological* status of reality in itself. In short, reality should be deterministically conceived in order to be fully knowable—at least in principle, although such a knowledge is never actually complete:

> Perhaps the metaphysical choice for global determinism is not particularly interesting for science […] In the never ending adventure of scientific research, one has of course to stop […] but such stops are due to the failures of our intelligence rather than to an 'essential' impossibility to go beyond. (Thom 1990b, pp. 277–278)

Thus Thom is assured against any possible charge of adhering to a sort of ingenuous pre-Kantian deterministic ontology. And nevertheless, his statements concerning a purely 'local' determinism—which I have defined as 'epistemological' in contrast to ontological and/or metaphysical characterisations—appear inconsistent when compared with two quite symptomatic passages in the text. The first refers to Einstein's old joke: 'I am among those who do not believe that God plays dice' (Thom 1990b, p. 275); the second is a note: 'the conflict determinism/chance is the manifestation of an ontological preference either for the substance or for the attribute' (Thom 1990b, p. 275, 279 note iv). In light of Einstein's joke, this second notation assumes the unexpected form of an ontological stance. In fact, when he associates determinism with substance *against* chance and attributes, Thom renews his attack on 'the deconstructors of being, the detractors of order and cantors of chaos', which actually 'prefer statistics to determinism' (Thom 1990b, p. 279).[4] Here Thom seems to reaffirm—at least implicitly, and consistently with his previous essay—an ethico-ontological choice for 'substance' and 'cause' as the basic tools of determinism against the 'popular epistemology' which would undo being into mere relations thus 'outrageously glorifying chance' (Thom 1990a, p. 61).

In this approach, substantialism and determinism are strictly entangled, within a kind of 'spontaneous philosophy of scientists' which returns to a remarkably

[4] In Thom's view the use of statistics is the mark of the impossibility to gain a complete deterministic description of a process (Thom 1990a, p. 66), and nevertheless it proves a valid hermeneutic function (Thom 1990b, p. 274).

ancient tradition.[5] According to my hypothesis, behind any substantialistic and deterministic epistemology one invariably finds a fundamental Cartesian-like metaphysical dualism. Since its Cartesian origin, this choice for determinism seems to maintain the burden of a metaphysical dualism which no phenomenological *epochē* can escape. Neither (neo)Kantian epistemology nor Empiriocriticism can be safely placed outside the perspective of a science imagining reality as a totalised complete system under the disincarnated look of the subject. This is quite evident in the implications of Laplace's 'radical' mechanism. As Thom himself recalled in his preface to the *Essai philosophique sur les probabilités*, although he denied any substantiality to the subject of scientific knowledge, Laplace could not avoid constructing—*such as* Descartes had done—a 'metaphysical-theological hypostasis' on which he could ground the postulate of exteriority of the subject's look in relation of its object-universe (Thom 1986, pp. 22–23).[6] It is worth noting that, in addressing this very criticism to Laplace, Thom in fact seems to collocate himself in a tradition according to which the 'arbitrary' institution of the subject of science depends on an 'ontological difference' between human being and nature: 'in order to assume epistemic significance, determinism necessarily requires human free will' (Thom 1990b, p. 272). An alleged ontological difference reveals itself here as grounded on a presupposed anthropological difference, a basic epistemological 'value' deeply-rooted in a supposed 'human nature', which would express (and celebrate) itself in the asymptotic, progressive tendency towards the research of truth:

> All these efforts in the search for truth tend to lead it [*l'esprit humain*] back continually to the vast intelligence which we have just mentioned, but from which it will always remain infinitely removed. This tendency, peculiar to the human race, is that which renders it superior to animals; and their progress in this respect distinguishes nations and ages and constitutes their true glory. (Laplace 1814, pp. 3–4)

Now, if it were true that Thom's morphodynamic structuralism would run the risk to develop—in Jean Petitot's words—'a neo-mechanism' (Petitot 1975, pp. 145–146), Simondon's criticism of determinism might be easily extended to it. And nevertheless my aim is not to reduce Thom's thought to the simplistic image of modernity I just evoked, but rather to explore the risky implications of the approach I named 'epistemological determinism'.[7] This allows me to sketch the ideal-type of modern mechanistic philosophies, as far as they refute any ontological value to the concept of chance, assuming it only as the limit-case indicating the shortage of

---

[5] I am borrowing the expression used by Althusser (1974). I assume here that the 'spontaneous philosophy of scientists' fails to grasp the differential relationship between science and ideology, in fact ideologically taking the scientific-mathematical representation of reality for reality itself. On Althusser's course, see Macherey (2009).

[6] Partially quoted in Prigogine-Stengers (1990), p. 248.

[7] Indeed, I am adopting the expression 'epistemological determinism' following Bachelard's suggestion of taking it in a sense deprived of any metaphysical presumptions (Bachelard 1951, p. 223), which differs from the meaning attributed to it by Bouquiaux when commenting the same *querelle du déterminisme* (Bouquiaux 1994, pp. 94–96).

complete knowledge of causes, which would be the essential postulate of scientific method. My enquiry concerns the possible discovery, beyond apparently different forms of determinism, of a similar theoretical structure which would prove an implicit adhesion to a dualistic metaphysics, i.e. an anthropocentric stance that de facto preserves for human beings a privileged place *in* or *out of* nature. In the present article I will try to show how Simondon's theory of individuation can allow us to move in a different direction, and develop a conjoint criticism of substantialism and determinism by endorsing another interpretation of the significance of the apparent 'distance' between human being and nature, and of the early-modern epistemological divide between knowledge and reality.

## 1.3 The Philosophy of Individuation and Its Paradigms

Simondon's philosophy of 'individuation', or of 'ontogenesis', is fully inscribed within the modern horizon of the philosophical system, as well as the cybernetic project which he assumes as a model when he is writing his doctoral thesis, *L'individuation*, during the 1950s. Simondon's aim is to provide relevant contribution to the epistemology of natural and social sciences, thanks to a theory of processes of information exchange at all levels, a model he transposes onto the different regimes of individuation: physical, biological and psychical-collective.[8] In fact, in all the different epistemological domains, an unquestioned and 'obscured zone' prevents the knowledge of ontogenesis. In particular, a substantialistic conception of the individual is what contributes to hide the processes of individuation, and it is therefore the main target of Simondon's attack to classical substantialism.

Thus *Individuation* starts with a double critique: on the one hand of the Aristotelian hylomorphic dualism of matter and form, on the other of the monistic reduction of nature to a fundamental substance (ILFI 23). Simondon's effort is, in both cases, to demonstrate the inadequacy of the conceptual apparatus of classic philosophy with regard to the results of twentieth century scientific thought, mainly quantum physics. For this reason, if it is true that the term 'individual' spans all the domains that *Individuation* ascribes to 'being', it is also true that Simondon distances himself from its classical link with the concepts of 'substance', 'essence', and 'form'. Furthermore, it is quite clear that only a redefinition of the concept of 'individual' could reveal what the philosophical imagination of a substantialised individual has always been hiding, i.e. the processes of individuation: 'to be rigorous, one should not speak of individual, but rather of individuation' (ILFI 191).

In two early programmatic texts, Simondon challenges directly the problem of the theoretical status of the philosophy of individuation, levering on the concepts of

---

[8] An extension likely to prompt the further observation of Canguilhem, his *directeur de thèse*, according to whom 'From the philosophical point of view, it would be a question of a new kind of Aristotelianism, on the condition, of course, that Aristotelian psychobiology and the modern technology of transmission not be confused' (Canguilhem 1943, pp. 277–278).

'structure' and 'operation (i.e. 'process').[9] While 'structures' are easily recognisable as the objects of existing sciences, in order to challenge the difficult goal of understanding 'operations' or 'processes', Simondon opts to refer to two 'basic intuitions' which should function as paradigms for the explication: crystallisation and modulation. The opposition between the two kinds of processes—which in fact poses a lot of hermeneutic problems—is nevertheless a good starting point to cross two fundamental and complementary themes traversing *Individuation*: (a) the theme of the double nature of the structured individual, considered both a system and *part* of a system, and (b) the theme of the mixed causality characterising processes.

### *1.3.1 The Structure (the Individual as a System)*

Modulation and crystallisation are for Simondon two different ways of understanding and describing the same processes at different levels, thus delineating a different representation of the individual depending on the level at which it is considered. In *crystallisation* the individual is understood as *a part* of a process which goes from the encounter between a simple individual (the crystal seed) and a milieu full of potentials (the supersaturated solution) producing a partially individuated system. Such an encounter functions as the trigger [*amorce*] of the system phase-shift [*déphasage*] into a complex individual (the crystal) and a milieu deprived of potentials (the low-concentration solution). On the contrary, in *modulation* the individual itself is understood as a metastable system made of different 'phases', the result of a coupling of initially independent systems and 'processes of formation' [*prise de forme*] which the hylomorphic scheme improperly divides into 'form' and 'matter', as in Simondon's example of the moulding of a brick (ILFI 40 ff.).[10]

It is precisely this double characterisation of the system that induces Simondon to introduce the concept of 'metastable system'. The classic 'stable' individual,

[9] Simondon uses there the term 'operation' as a synonym for 'process'. Although they first appeared undated in the second edition of *L'individu et sa genèse physico-biologique* (1995), the two programmatic texts *Analyse des critères de l'individualité* (in ILFI 553–558) and *Allagmatique* (in ILFI 559–566) clearly express Simondon's need for an overall view on the project of *Individuation* which he probably sketched before or during its elaboration (1957–1958). They could also be a further revision of its outcomes, in view of the 'general theory of social sciences' Simondon exposes in his paper at the *Société Philosophique* in 1960 (in ILFI 531–551). However, in this case the absence of the concept of 'transductive operation', still central in Simondon's paper, could hardly be explained.

[10] Simondon borrows the term 'phase' from physics and chemistry in order to indicate how different processes, parallel, divergent or convergent, are simultaneously taking place in a system. In short, the physical notion of phase serves him to undermine a substantialistic representation of the individual, conceiving it as a 'phase shifted' system simultaneously crossed by different and divergent processes. The individual itself is thus a system made of phases and thresholds which can put different systems in relation.

identical to itself, becomes in this sense the impossible limit-case of a perfectly static system, the fictive name for a completely accomplished process of individuation, while in actual fact one is always witnessing processes of individuation that deprive individuals of any fixed identity:

> The relation of being with itself is infinitely richer than identity. Identity, a poor relation, is the only relation of being to itself that one can conceive according to a doctrine which considers being as single phased. (ILFI 318)

The difficulty of Simondon's attempt to conceptually dismantle the traditional ontology of identity, is evidenced here by the ambiguity the term individual retains throughout the whole text of *Individuation.* This ambiguity carries on a double meaning: the one prevalent within the philosophical tradition, the individual to which Simondon refers as the structured 'part' of a system in the course of individuation, and the individual crossed by or, better, emerging from processes taking place at different levels, to which Simondon often implicitly refers as a system in the course of individuation. It is precisely through the concept of 'metastable system' that Simondon refers to a being which is 'more than a unity and more than an identity' (ILFI 26).

### *1.3.2 The Process (Individuation as Operation)*

Although a mechanical sequence, the beginning of the *process of crystallisation* is irreducible to the sequence itself. The process is triggered by the encounter of the system with the singularity of a crystal seed: an encounter which cannot be strictly reduced to the sequence it triggers, and therefore is not determinable within the system itself.[11] On the other hand, the *process of modulation* begins when different systems converge. Such a process could be considered determinable only at the level of the accomplished (macro) system, where in fact there would be no *emergence* of a new system, but merely the assemblage of two subsystems.[12]

[11] Whether it is introduced from the outside or emerging from a causal encounter of molecules: 'A seed crystal can be replaced in certain cases by chance encounters, i.e. by a chance correlation between molecules' (ILFI 550).

[12] One must at all costs avoid any interpretation of the relations among different scale systems as a kind of Chinese box game culminating into a Nature-whole conceived as a System including all systems, since this is exactly what Simondon explicitly denies when challenging Kurt Goldstein's 'Parmenidean ontology' and asserting his own theory of systems as metastable, phase-shifted and 'in state of disparation', therefore incomplete and not entirely determinate (ILFI 229). According to Simondon 'Nature' conceived as a macro-individual would be the silent and perfectly stable—dead—universe of maximum entropy; on the contrary, we are exclusively concerned with 'non-totalised' systems: 'Systems cannot be *totalised*, since the fact of considering them the sum of their elements spoils the awareness of what actually makes them systems: relative separation of the sets it contains, analogical structure, disparation in general, relational activity of information' (ILFI 234, n. 1). Goldstein's book *The Organism* (in the German original: *Der Aufbau*—the Structure—*des Organismus* 1934) is a Gestalt-like approach to organism through a joint study of biology,

Simondon's original hypothesis was that each crystallisation is in fact a reversed modulation, and vice versa (ILFI 566). But what the two 'paradigms' of modulation and crystallisation really share as processes, is the fact that none of them can be entirely reduced to a deterministic sequence of cause-effect relations.

In *Individuation*—according to the inspiring methodological paradigm of quantum physics—all processes are characterised by a fundamental discontinuity and by reiterated changes of the order of magnitude. And on this topic it is worth recalling how Simondon's debt to the physicist Louis De Broglie, even if not always evident, is constant and decisive throughout *Individuation*.[13] Although referring to microphysics, the discovery of the 'indeterminacy principle'[14] poses philosophical problems concerning not only the deterministic characterisation of classical mechanics, but also the status of all sciences related to objects of a different magnitude in which, however invisible and non explicitly described, such factors still produce effects. De Broglie's argument entails the possible expansion of the approach derived from microphysical discoveries, first of all to biology,[15] and, furthermore, to social sciences: 'its [microphysics] relevance is not limited to the domain of physical sciences, it applies to the sciences studying life, human beings and human societies' (De Broglie 1947, p. 225).[16] In this perspective, at all levels, from microphysics to social systems, for each process there are both determined conditions of state (i.e. possible effects and impossible ones), and indeterminacy margins of becoming, which exclude any uniform, linear and continuous relation between causes and effects. In Simondon's view, a conception of the individual as a metastable system involves a complete overhaul of the methodology of the social sciences, and a task comparable to the one which the natural sciences seemed to be achieving through questioning the ontological status of their object:

> Could we do the same in the social sciences? Could we found a social science [*la Science humaine*] respecting, of course, multiple possibilities of application but having at least one common axiomatic applicable to different areas? (ILFI 355)

(Footnote 12 continued)
psychiatry and medicine, which had great relevance for an entire generation of French philosophers during and after the Second World War.

[13] The importance of De Broglie is is quite clear if one considers that, among the only 20 bibliographical references in *Individuation*, three are De Broglie's (*Archives de Georges Canguilhem*. Paris ENS, CAPHES: GC. 40.2.1).

[14] Even if the current English translation is 'uncertainty principle', the original term used by Heisenberg was *Unbestimmtheit*, which can also mean 'indeterminacy'. I will use the second term, since it better expresses an ontological lack of determinacy rather than an epistemological uncertainty of knowledge.

[15] 'A mammal, for instance, belongs to the microscopic world as far as the elements directing its living dynamics are collocated, in effect, at the level of atomic systems. The functioning of living systems will therefore be studied 1 day thanks to microphysical concepts' (De Broglie 1947, p. 162).

[16] On the philosophical relevance of early twentieth century microphysics, see in particular Chap. 7 on *Les révélations de la microphysique*, and Chap. 11 on *Hasard et contingence en physique quantique*.

In short, if all processes have a determinate tendency due to their irreversibility, nevertheless the actual processes can never be deduced from the initial state of the system. This perspective excludes an exhaustively predictive science. It, on the contrary, necessitates a 'double' theory, concerned on the one hand with conditions of state and structural tendencies and, on the other, with the ontogenesis of the *singular* operations of individuation. Simondon refers continuously to such a theory in the programmatic texts and in *Individuation* with the peculiar name 'allagmatics' [*allagmatique*]' (ILFI 565–566). For him, an adequate theory of individuation cannot rely on a purely topological science as well as on a 'pure' science of the operations of individuation; since both concern limit cases, they abstract from the real process and are therefore singularly incapable of giving an adequate account of it. For this reason in *Individuation* allagmatics finally appears as a 'theory of metastabilities', i.e. of 'the exchange processes between spatial configurations and temporal sequences' (ILFI 238). In his mind, since the individual is made of processes, the sciences of individuals must open to a theory of the 'operation of individuation' which cannot be a topological theory.

### 1.3.3 Singularity, Transduction and the Individuation of Knowledge

In *Individuation* Simondon seems quite clear about his way of conceiving 'topology', when he states: 'What a topology lacks is the consideration of potentials. Precisely because they are not structures, potentials cannot be represented as graphical elements of the situation' (ILFI 238). Thom harshly criticised Simondon, because of his presupposition of the primacy of discontinuity, while, on the contrary, 'the continuous is *ontologically anterior* to the discrete' (Thom 1994, p. 105). Furthermore, he claimed that—due to inaccurate knowledge of topology—Simondon had failed to develop an adequate analysis of the 'subject of transductive knowledge', and subsequently he could not provide a satisfying theory of signification, and this error entailed his overly fluctuating [*mobiliste*] description of the principle of individuation (Thom 1994, p. 105, 110–111). Unfortunately, in his essay Thom—who ironically complained about the obscurities of the philosophical jargon in *L'individu et sa genèse physico-biologique*—did not even mention the last part of *Individuation*[17] where, on the contrary, Simondon provided a theory of

[17] In 1964 Simondon published *L'individu et sa genèse physico-biologique* which contains the introduction, conclusion and the first two parts of *Individuation*, with the exception of chap. I. 3. Only in 1989 did he re-edit *L'individuation psychique et collective. À la lumière des notions de forme, information, potentiel et métastabilité* (published posthumously), which contains the conclusive part on psychical and collective individuation, a new introduction made up of the original one plus the paper he delivered at the *Société Française de Philosophie* in 1960 (*Forme, information, potentiels*), and the unedited *Note complémentaire sur les conséquences de la notion d'individuation*. In a 1995 edition the two programmatic texts *Allagmatique* and *Analyse des*

signification which might be actually related to Thom's semiotics, as Petitot suggests when he includes Thom's morphodynamics and semiophysics among Simondon's most surprising scientific anticipations (Petitot 2004, pp. 104–106).

Despite his severe critique, in his essay Thom grasps actual problems in Simondon's treatment of topology: in his work there is no concern for quantitative topology. One can understand the qualitative sense Simondon attributes to the term 'topology' when—in the entry 'Modern Psychology', co-edited by Simondon for the *Encyclopédie de la Pléiade* shortly before writing *Individuation*—he refers to K. Lewin who, in *Principles of topological psychology* (1935), 'proposes a theory of the psychological field grounded on the notion of non-quantitative space of topological geometry' (Simondon 1957, p. 1676). And nevertheless it is likely that from the 1970s Simondon might have seen in Thom's topology of linguistic and biological systems the same universalising potentiality he had attributed a decade before to the concept of information elaborated by cybernetics. *Individuation* itself was in fact intended to be a huge reassessment of the concept of information, in order to make it 'non-deterministic'. Simondon inherited the concept of information from one of the fathers of cybernetics, Norbert Wiener, who conceived information as a paradigm which could be at least in principle extended to all the fields of scientific research: biology, psychology, psychopathology, sociology and political economy. In that period Simondon was adopting the same key methodological tools for his philosophy:

> The language of cybernetics, already applicable to nervous system physiology, could prove to be suitable for describing the relations between the human being and his natural and social milieu, overcoming the alternative between liberty and determinism, which seems to be the major obstacle for any psychological science. (Simondon 1957, p. 1701)

Indeed, Simondon considered the notion of information, elaborated by cybernetics in connection with the concept of homeostasis, insufficient for explaining the operating of complex systems. Therefore it had be reformulated in terms of a differential relation and 'operation'. Thus he felt confident that a critical enquiry into the concept of information could supply a paradigm to direct his own quest for a science of the relations between structure and operation.

This is what he recursively did through the whole of *Individuation*. The different series of problems Simondon's research encounters there, are clearly embodied in his subsequent attempts to define different 'paradigms' of operation. Crystallisation and modulation are in fact two of them, but in *Individuation* Simondon introduces a new key concept by referring to an 'operation' which he calls 'transductive operation', 'transductive process' or more simply 'transduction'.[18] The attempt to

---

(Footnote 17 continued)
*critères de l'individualité* also appeared. Only in 2005 all these texts were gathered together and published as ILFI, with the further addition of the unedited *Histoire de la notion d'individu*, originally supposed to be the second part of *Individuation* which Simondon never completed.

[18] The concept of 'transduction' has origins both biological and technological, and refers to a mode of propagation—a non-deterministic sequence, presenting possible gaps and discontinuities.

convey the two former paradigms within a unique notion shows how in *Individuation* Simondon definitively abandoned (if he ever had it) the assumption that modulation and crystallisation could describe two 'essentially' different kinds of processes, and rather used the notions to describe different and concurrent aspects of the same processes at different levels.[19] The categorisation of such paradigms was further widened and complicated by Simondon when, a couple of years later, he proposed a new synthesis of 'continuous modulation' and 'discontinuous transduction', clearly having biological connotations, of 'organising amplification' (Simondon 1962, p. 417).[20] The same inexhaustible quest for an 'analogy of dynamical situations' is what most likely pushed Simondon to attend Thom's seminar during the 1980s, even though the latter's work on differential topology dated back to the early 1950s.

What is most interesting for the purpose of this article, is to discern what actually links all of Simondon's recurrent *attempts* to 'name' the processes of the discontinuous propagation of information and reconfiguration of metastable structures. A key is offered by the concept of 'singularity'. In effect, Simondon builds the concept of transduction *against* the cybernetic concept of information and *beyond* the implicit holism of *Gestalttheorie.*[21] Transductive sequences are triggered by singularities which are […] 'historical and local' (ILFI 81). Now, this coupling of the terms 'singularity' and 'historicity', is quite interesting, since it functions at all levels (or 'regimes') of individuation, from human societies to matter. As a striking evidence of that, it is worth recalling the way the concepts of singularity and historicity extend over all the fields of 'being' Simondon defines: at a physical level there are 'historical singularities brought about [*apportées*] by matter' (ILFI 57); in crystallisation 'there is therefore an historical issue in the occurrence of a structure in a substance: the structural germ has to appear' (ILFI 79), and 'the individuation of an allotropic form starts from a singularity of historical nature' (ILFI 80); at the biological level 'the individualisation of the living being is its real historicity' (ILFI

[19] Those who would claim that in *Individuation* the paradigm of crystallisation exhausts the significance of transduction by quoting the paragraph *Individuation as a genesis of crystalline forms from an amorphous state*, would end up in quite a bit of difficulty when trying to justify the following statement: 'individuation is a modulation' (ILFI 220).

[20] When editing the acts of the conference, Simondon decided to substitute it with this brief abstract, while the entire paper was recently published as *L'amplification dans les processus d'information* (Simondon 2010, pp. 157–176). In July 1962 the prestigious *Colloque international de philosophie* traditionally held at the Royaumont Abbey was dedicated to *The Concept of Information in Contemporary Science*. Simondon's paper clearly represented for him a further occasion to reformulate an old problem. The theory of information, such as he elaborated it in *Individuation* as a relation between metastable systems and the 'incident signal', underwent no substantial modifications at Royaumont, where it provided the common basis for the three paradigms of 'amplification'.

[21] For Simondon the validity of the 'postulate of isomorphism' is rather to be found in morphogenetic processes (i.e. individuation processes), since the holism of 'form' does not in itself escape a deterministic horizon (Simondon 1966–1967, p. 298).

268); at the psychical level 'we believe any thought, precisely as far as it is real, is a *relation*, i.e. it entails an historical aspect in its genesis' (ILFI 84).[22]

In short, the two terms appear as quasi-synonymous and point to the same double aim: on the one hand, they prevent any attempt to safely circumscribe the place of humans within nature, and, on the other hand, they show the extensions of the same common properties of processes—discontinuity and non-reversibility—to all degrees of being and knowledge. Furthermore, this common characterisation of all processes entails the impossibility of providing any exhaustive axiomatisation of scientific research through the conceptual framework of sciences which postulate the dependence of reality on the principles of their epistemology and the models one can derive from it, and, in the last instance, from their transcendental or metaphysical support, the subject of science. In conclusion, the effort of philosophical reflection would be therefore unanchored from any 'anthropological difference' which might grant its 'divine' theoretical distance from scientific research. Philosophy rather emerges from scientific research itself as a real ontogenetic process taking place at the level of the bodies of knowledge individuating as symbolic systems:

> Therefore it is neither an immediate knowledge, nor a mediate one, that we can have of individuation, but rather an operation of knowledge which is parallel to the known operation; we cannot, in the common use of the term—*know individuation*; we can only individuate, individuate ourselves, and individuate in ourselves [...] Beings can be known through the subject's knowledge, but individuation of beings cannot be grasped out of the individuation of the subject's knowledge. (ILFI 36)

Precisely because thought proceeds transductively, it cannot be consistently formalised through the classical principles of identity and the excluded third (ILFI 324). More crucially, a logic in the classical sense is impossible. The problem can be posed in the following terms: if the 'transductive operation' is singular—i.e. its origin, course and results cannot be subsumed under any universal concept—then the foundation of *a single* logic of transduction will be impossible by definition. In fact a 'theory of being preceding any logic' according to which there are 'multiple kinds of individuation', should rather produce 'multiple logics, each corresponding to a definite kind of individuation' (ILFI 36), thus resulting in the practice of a

---

[22] Given the importance of the concept of 'singularity' in Deleuze's philosophy and the way he draws on Simondon's terms since his 1966 review to *L'individu et sa genèse physico-biologique*, a clarification is needed concerning the different usage. For both philosophers the concept of 'singularity' points to a discontinuity concerning processes, but they differently conceive the relation between 'singularity' and 'individual'. Deleuze collocates the individual on a different scale (molar) in relation to the pre-individual regime of singularity-events (molecular) his 'transcendental-empiricism' is concerned with. On the contrary, for Simondon's philosophy of individuation, the individual is to be understood as a part of a discontinuous process *without reducing it* to a kind of epiphenomenon of molecular features. Indeed, Simondon's use of the terms 'singular' or 'singularity' is a very restricted one, which refers to a structured individual *when* it is the 'seed' or the result of a process triggered from an aleatory encounter. Transductive processes are aleatory precisely due to this 'historical' aspect in their genesis: the presence of what one could define as an 'evental singularity' [*singularité événementielle*].

'pluralism of individuation' (Simondon 1989, p. 217). This kind of logic of singularity would incorporate and repeat the aleatory factors of the transductive process itself, and nevertheless the elevation of transduction to the role of a methodological paradigm does not entail that any *formal* logic can be simply deduced from the ontological assumption of transduction. That is how Simondon defends the validity of analogical thought: *grounded* on ontological transduction[23] but not *guaranteed* nor *established* by any possible logical formalisation. It should be quite clear that such a 'clinical' and possibly 'genealogical' perspective entails the definitive abandonment of the early-modern perspective, both of mechanistic determinism and of its epistemological correlate: the disembodied look of the subject of science.

In this light it should be clear how Simondon's former reference to the paradigms of modulation and crystallisation served a double strategic issue: it was intended to contribute to the dismantling of the assumption of the dominance of the cause-effect relations as the ground for ontological determinism, and, therefore, to prevent the understanding of the individual as a substance identical-to-itself. Furthermore, *Individuation* itself appears as a powerful experimental-setting of concepts, where Simondon tests his philosophy of individuation through the analogical transposition of different paradigms, a philosophical 'operation' he always intended to maintain current with contemporaneous technical and scientific research:

> Such a representation of being necessitates a conceptual reforming which depends on a radical revision of some basic schemes, some new paradigms which would replace the hylomorphic scheme imposed by culture [...] We have tried to derive a paradigm from physical sciences. (ILFI 319)

Thus, the genuinely philosophical performance of *Individuation* lies in the attempt to enable different 'schemes'—modulation, crystallisation, phase, metastability, transduction, and many other conceptual tools—for the analysis of the different domains of being, of their structural conditions and operational status: tools for defining the thresholds between different domains in order to make a problem of them, rather than to fix them. Simondon's concepts, in fact, do not define any separate 'realms'—matter, living beings, psyche, society—traversed by individuals or any kind of substances of which individuals would be composed. On the contrary, they indicate 'phases', processes, whose dynamic composition continuously constitutes and modifies the configuration of individuals, as it happens within a magnetic and gravitational field, in which different forces and processes constitute an irregular and unstable space, full of potentials, which can modify or be modified by whatever—matter or energy—becomes part of it. No ontological guarantee, then, of a stable and secure domain, and no science capable of defining the specific processes characterising a domain, is possible without a prior enquiry of the singular ontogenesis and functioning of the different structures which constitute it. It becomes clear then that Simondon's philosophy of individuation is based on

---

[23] 'The possibility of using an analogical transduction for thinking a domain of reality, indicates that such domain is the actual place of a transductive structuration' (ILFI 33).

the critical analysis of the results of the sciences of structures, aiming to criticize rather than to confirm the alleged 'identity' of their objects, thus reactivating the ontogenetic hypothesis in order to discover the actual tensions which render each structure metastable, i.e. an individual undergoing individuation. This explains the path followed by his analysis: the two sections of *Individuation* display the process of individuation through three different levels (physical, biological, psychical and collective)[24]; and each 'movement' concludes with the demonstration of the insufficiency of any structural and topological definition of the individual which would not adequately consider the problem of time.

The encyclopaedic extension of *Individuation* gathers all the materials for a unified theory of the sciences; the wide breadth of this study progressively transfers the ontological assumption of the quantum paradigm into the epistemological creation of a 'non-deterministic' theory of information, in order to provide an 'axiomatic for the social sciences'. In fact, with his frontal attack on substantialism on the basis of a non-deterministic conception of *all* processes, Simondon aims to reconstruct the field of social sciences providing a new collocation for classical dualisms (body/mind, immanence/transcendence, necessity/liberty) within ontogenesis. This epistemological production of compatibility is precisely what he usually calls 'axiomatisation'.[25]

According to this view, *Individuation* looks like a huge encyclopaedic work which, starting from the 'facts' and models of sciences, *experiments* with the philosophical possibility of unfolding their ontological and epistemological presuppositions. But Simondon's undertaking is twofold: on the one hand it emerges as the repetition of this philosophical attempt next to any term, concept and image he draws from each field of scientific research, and on the other it repeatedly collapses into a systematic, totalising project. For this reason the problem of a theory of ontogenesis in *Individuation* both recurs under the imaginary shape of a dream of axiomatisation, and appears as a genuinely philosophical enterprise. The philosophical force of Simondon's oeuvre, of which he is not always aware, emerges in the obstinate repetition of the same operation of structuring a subject-object relation (i.e. of knowledge) at the exact scale of each of the systems concerned. It is an irremediably singular operation about which he at least once explicitly assumes the impossibility of providing a definitive formalisation:

[24] It is worth noting that in Simondon's original thesis the 'macro-domains' of individuation are only two, since the psychic-collective one is part of the individuation of living-beings. But this does not foil my argument, because it is true that the descriptions of individuation at every domain *and* sub-domain follow the same pattern. Simondon's original thesis can be consulted at the *Archives de Georges Canguilhem*, GC. 40.2.1, ENS Paris, CAPHES.

[25] According to Guchet, Simondon's program of an axiomatic of the social sciences must be collocated equally distant from a 'scientific positivism' and a phenomenology of the immediate access to proto-experience. This program would link Simondon and Merleau-Ponty to the aim of 'reaching the concrete human being, starting from positive knowledge (psychology, sociology, history)' (Guchet 2001, p. 103). Although Guchet's analysis tends in a way to explain Simondon against the backdrop of Merleau-Ponty's work, it shows some important points of contact between the two which are worth taking into account.

> It might be that ontogenesis cannot be axiomatised, which would explain the existence of philosophical thought as perpetually marginal in relation to all the other studies. Philosophy would be the kind of thought set in motion by the implicit or explicit research of ontogenesis in all orders of reality. (ILFI 229)

The singularity of individuation, in short, requires equally a singular exercise of *knowledge*, the kind of 'clinical' thought Simondon defined, a few years later, as a regime of functioning in which 'the appropriation of each problem is perhaps the higher goal philosophy can devote its efforts to' (Simondon 1965, p. 16).

### *1.3.4 A Departure from the 'Deterministic Age' of Metaphysical Dualism*

In his course *Initiation à la psychologie moderne* (1966–1967), Simondon characterised the 'deterministic age' as the one which postulated the order of Nature as 'uniform, necessary, universal and analytical', i.e. continual, deterministic, general and reducible to elementary elements. According to Simondon the 'deterministic age' began to collapse at the end of nineteenth century, initially attacked by evolutionary biology, then by holistic assumptions based on Maxwell's theory of fields (Simondon 1966–1967, pp. 288–290). From Simondon's perspective—as in the eyes of many scientists and philosophers since the beginning of the last century—quantum physics seemed to present unprecedented evidence: matter was in no way reducible to imagination, not even to a possible mathematical kind, and certainly not the one which classical mechanics had dreamt up. It is in the light of that *thaumaston* that I am now reading Simondon's effort to provide a philosophy of individuation. Simondon's *Individuation* can thus be interpreted as an attempt to overcome the very horizon that grounded both the determinism and the substantialism of modern philosophical mechanism: metaphysical dualism. His philosophy of individuation was intended to contribute to the dismantling of both the modern image of nature as a deterministic machine, and its complementary onto-theological image: the transcendence of reason. In effect, he could neither admit any 'Other' from nature to which human beings would somehow pertain (in the different forms it was depicted, from the Cartesian *Res cogitans* until the Kantian *Ich denke*), nor could he consider the conception of singularity he derived from quantum mechanics compatible with the Cartesian conception of *Res extensa*:

> A particle which can be represented as the singularity of a field cannot be conceived within Cartesian geometry, because one cannot introduce singularities in the space of *Res extensa*, extended substance, without modifying Cartesian geometry and mechanics. (ILFI 144)

Rightly or wrongly inspired by the paradigmatic power of quantum physics, in *Individuation* Simondon carried out research he had already set in motion in his programmatic writings, where, through an explicit declaration of intent, he linked his philosophical project to a reform of the concepts of individual and of cause:

> At the end of this double study [concerning the concepts of modulation and crystallisation], the philosophical notion of *causality* will be enriched and the notion of individual defined. (ILFI 566)

It is from this very perspective that determinism shows its original connection with substantialism. The two concepts originally derived from the same ideological matrix, which postulated (and in fact imagined) an ontological gap between the order of reality and the order of knowledge, and therefore a kind of isolation: human thought as an exception within the realm of nature. This operation had promptly been condemned by Spinoza, who intended to use determinism as a conceptual lever in order to scatter Aristotelian and Scholastic teleology.[26] Nevertheless, as it emerged in Cartesian mechanism, determinism assumed the paradigmatic function of isolating the human species from its milieu, by reclaiming its straight metaphysical substantialisation.[27]

According to Prigogine and Stengers it was not quantum mechanics, but thermodynamics that marked the beginning of a possible 'new alliance' between human beings and nature, long after the early-modern breakthrough. The study of dissipative structures and of processes far from equilibrium might possibly end the faith in the modern myth of a science 'capable of discovering *global* truths about nature' (Prigogine-Stengers 1985, p. 44, 51). That myth had had its apex in the metaphysical dualism condensed in Laplace's *demon* and still lasts in the 'unwittingly repetition of the ritual of the ancient faith' (Prigogine-Stengers 1985, pp. 75–77). In Stengers' view, Simondon is fully inscribed into this mythical tradition, since in *Individuation* he ceased to grasp the 'properly *technical* characterisation of the physical-mathematical notions he tried to generalise'. For this reason she invites us to resist the seductions of a philosophy of nature present in Simondon's work as something 'one and undivided beyond specialised knowledge' (Stengers 2002, pp. 308–309). This is exactly what I have attempted to do in this essay by underlining what in Simondon's own thought might lead, on the contrary, towards a non-totalising technical and clinical exercise of philosophical thought.

In the above mentioned debate, Prigogine and Stengers also reacted to Thom's essay with *La querelle du déterminisme, six ans après* (1990). In this work the authors opposed two different but convergent ways to conceive the relation between science and reality, respectively pictured through 'Laplace's *demon*' and the 'Augustinian—Leibnizian god'. Laplace's demon would embody the ideal physicist, who knows all the elements of the state of a system and can therefore exhaustively foresee its future transformations. This stance corresponds to a false omnipotence, a science capable of admitting its limitations only dependent on a lack of empirical experimentation, or on scarcely elaborated information. On the

---

[26] See Spinoza's well-known critique, *within* the deterministic tradition, of the prejudice of those who imagine the existence of an '*imperium in imperio*' (*Ethica*, Pars Tertia De Affectibus, Praefatio).

[27] About the grounding function of Descartes' metaphysics for his physics, see Garber (1992, 2001).

contrary, the Augustinian god's absolute infinity and detachment from natural laws, marks the absolute distance of any possible humanly-constructed logic or model from reality. This stance would entail the limitation not only of the deterministic model, but of the objective value of science itself, conceived as a form of knowledge based on the concept of substance, against which Stengers shifts the focus onto the concept of relation.

Of course this would still appear unfaithful to a radical 'epistemological determinism', and even a *false alternative* to determinism, if it authorised the assumption of an 'ontological' primacy of relations. Indeed, Simondon avoided opposing substantialism to a realism of relations *tout court* which would entirely dissolve being in the 'substantiality' of relations, and eventually culminate in what he called a 'Parmenidean'[28] conception of a continuous whole, i.e. an equally substantialistic metaphysical assumption. It is in this sense that for Simondon 'dualism is still too monist in order to be maintained; it implies a substantialism' (ILFI 553). He considered the two errors specular, as far as they failed to exit from the false alternative imposed by the ideological assumption of substantialism: an alternative he openly refuted when, discussing the double solution theory in wave mechanics, he claimed that 'the individual is *being and relation*' (ILFI 143).[29] Simondon's philosophy of individuation resulted in the end in a kind of open discontinuity, which 'seems to be connected to the quantic issues which appear in physics' and 'presupposes a concatenation of physical reality up to superior biological forms, without establishing any distinctions between classes and genres' (ILFI 158).

Now, this 'theory of nature' implicit in Simondon's theory of individuation cannot be understood as a 'pre-Kantian' ontology, it is rather a move beyond the mechanistic and deterministic universe which not only roots the modern metaphysical divide but, furthermore, still haunts the phenomenological postulate of a reassuring presence of sense in discourse and *therefore* in nature.[30] This move strongly rejects any form of idealism, even concealed, by avoiding a 'linguistic

---

[28] See above, note 12. In this sense Leibniz is the paramount author of substantialism (ILFI 454). In fact, Spinoza himself is interpreted by Simondon as a 'Parmenidean' thinker.

[29] If the entirety of Simondon scholars take for granted that from crystals to higher degrees of complexity one can speak of individuals, this is more problematic concerning the subatomic world. See for instance the discussion with Aspe and Bontems, thus concluded by Stengers quoting Simondon: 'It is unclear whether what we call a critical fissile mass is not an individual as such' (Stengers 2002, pp. 318–319). Simondon is frankly ambiguous on this topic, but I assume consistent with his thought that there are no limits of order or magnitude for individuality: thus with the photon one can have, 'synthesised in the same being, and carried by the same carrier, both a structured and an amorphous measure, a pure potential' (ILFI 102).

[30] This is the summary of Ricœur's accusation during the debate of Simondon's paper (1960): trying to 'construct the universe of discourse from the region of nature which is itself something included within discourse', Simondon's proposal would experience irremediable paralogisms. Simondon's reply sounds his refusal to assume the (transcendental) horizon of language, and in this perspective we can read his 'theory of nature' as a step beyond the phenomenological postulate of 'sense' as the original warranty for any kind of discourse and *therefore* for nature itself. Through an adequate theory of information, the radical disjunction of the issues of signification and language in a way postpones the problem of sense to an ontogenetic issue: 'There

turn' which might possibly conceive language or 'sense' as a 'transcendental horizon', as Simondon clearly states when countering Ricoeur in the discussion which followed his conference at the *Société Française de Philosophie*:

> How could one admit that nature is part of the discourse? This is the postulate underlying your argument, and this is what I shall definitely refuse. (Simondon 1960, p. 759)

From a completely different non-phenomenological perspective, Simondon attempts to re-read Merleau-Ponty's courses on nature that were delivered at the time *Individuation* was being composed (Merleau-Ponty 1995–1960). He attempts this re-reading devoid of any supposed predominance of the (transcendental) subject, yet nevertheless flirts with idealism. In what Simondon calls the 'individuation of knowledge', the subject is able to recall its own conditions of possibility and identifies them as the causes both of individuation in general and of its individualisation in particular:

> If knowledge can trace back the lines which allow for the interpretation of the world according to stable laws, it is not because in the subject some a priori forms of sensibility exist, the consistence of which with raw data derived from sensation would be inexplicable; this happens because being as a subject and being as an object come from the same primitive reality, and thought, which now seems to institute an inexplicable relation between the object and the subject, in effect continues the initial individuation. The *conditions of the possibility* of knowledge are the actual *causes of existence* of the individuated being. [...] It is because individuation is the universal ground of the relationship between the object and the subject, that knowledge can be universal. (ILFI 264)

This is the operation in which the subject simultaneously finds and produces, i.e. invents, the universal grounding for knowledge. In this sense Simondon's method is the subjective continuation of an actual transduction, which allows the risk of a singular solution. Evidently, the success of such a reflexive operation cannot be guaranteed by any methodological formalisation. What Simondon calls the 'analogical method' is in fact the method of invention, an operation both theoretical and practical, which lacks any guarantees, as Simondon himself reveals when defining his own philosophy as 'a dramatic theory of the becoming of being' (Simondon 1960, p. 755).

One can notice here the destructive 'fascination with *clinamen*' emerging again, and pushing philosophy over the boundaries of the scientific assumption of ontological determinism. Although in Simondon's philosophy chance is not at all an element for (pseudo)explanation, as already shown the concept of singularity plays a key role there. It marks the threshold of collapse not only for deterministic explanations, but also for the very interpretive framework that sustains the apparent opposition of determinism and chance itself. In this sense the concept of singularity deprives the *querelle* over determinism of its own ground:

(Footnote 30 continued)
is a theory of nature in what I tried to present, which could not admit such a theory of signification as included in the Word' (Simondon 1960, p. 760).

> It is possible, in the last instance, to suppose that the theory of singularity can be ascribed neither to the framework of a deterministic physics nor to the framework of an indeterministic physics. The two would rather be considered the particular cases of a new representation of the real that one might call the theory of transductive time or theory of the phases of being. This completely innovative mode of thinking—which conceives determinism and indeterminism as mere limit-cases—can be applied to different domains of reality beyond the one of elementary particles. (ILFI 144)[31]

## 1.4 Beyond the *Querelle* Over Determinism: On Canguilhem's Footprints

Following Simondon we are brought to the point of abandoning the modern mechanistic and deterministic perspectives, but also the *querelle* over determinism itself; it is necessary to ask, in a preventative fashion, whether his texts evoked the aforementioned Althusserian question: how can the endurance of determinism as a 'spontaneous philosophy of scientists'—i.e. as an ideology—be explained?[32] I will try to sketch a Simondonian critique of determinism by drawing from the model he uses to critique hylomorphism—in *Individuation* a conceptual synonym of 'substantialistic dualism'—and extend it to determinism itself by way of a brief detour through some of Simondon's sources.

Substantialism, according to Simondon, has its roots in a biological attitude of 'homo faber' towards technical manipulation. This is what he aims to explain when countering 'hylomorphism', which causes knowledge to be knowledge of individuals, instead of knowledge of processes of individuation. But his techno-biological critique only provides the natural basis, not the whole explanation of the substantialistic 'bias':

> If only the individual and technical operation existed, the hylomorphic scheme could not emerge […] What the hylomorphic scheme first of all shows, is a socialised representation of work and an equally socialised representation of the individual living being. (ILFI 50–51)

Simondon subsequently provides a more genuinely sociological argument concerning the historical success of the hylomorphic scheme, referring to the different relationship established by the master [*maître*] and the artisan [*artisan*] in regards to

---

[31] 'It is only abstractly that one can speak of an absolute indeterminism (realisable through a complete internal resonance) or an absolute determinism (realisable through a complete reciprocal independence of chronology and topology). The general situation is a certain level of correlation between the chronology and topology of a system' (ILFI 148). See Bontems commenting this passage in order to show how 'Simondon's theory of individuation puts in relation the complexity of physical systems with morphogenetic processes' (Bontems 2010, p. 89).

[32] This ideology is in fact—as such—not at all spontaneous, and its emergence has therefore to be explained: 'I have only one more word to say about this "spontaneous" ideology: we will see that it is "spontaneous" because it is not' (Althusser 1974, p. 35).

the technical operation. The master's abstract relation is that of property, while the artisan's is the concrete immersion in matter through the technical process of its transformation. This 'evocation' of the master-slave dialectic is less a Marxian debt than a Hegelian reference to the abstraction of the master's knowledge in front of the servant's ability to grasp the singularity, the 'implicit forms' of worked matter (ILFI 57–60).[33]

Determinism similarly fulfils first of all a 'techno-biological' task, which one could define as strategic, due to its perceptive efficacy in foreseeing and thus orienting action. This explanation assumes different forms within Simondon's possible sources. In De Broglie's *Les conceptions de la physique contemporaine et les idées de Bergson sur le temps et sur le mouvement*, one can find a reference to Bergson's note in *La pensée et le mouvant* (1934), according to which only at the 'macroscopic' level of the living beings' perception 'reigns the apparent determinism that makes their action on things possible' (De Broglie 1947, pp. 210–211).[34] In short, it is again *homo faber*, or rather organisms in general, who, through the deterministic perceptual filter, reduce complexity and shape their own milieu in view of possible actions. Piaget, who was also an important source for Simondon, in *La causalité physique chez l'enfant* (1927) individuated the evolution of this pattern in the changing relation between children and their milieu, which entails the progressive elimination of chance from nature (Piaget 1927, p. 310).[35] Only a few years later Kojève (1932), starting from the assumption that the causal relation is no more understandable than the statistical one, posed himself this question: 'why such a "sharp preference" for causality?' He looked for an answer in introspection, an internal experience in which the object of desire determines the human being in action: this teleological causality the human species would subsequently transfer to the external world (Kojève 1932, p. 123).

In fact, these explanations do not dismiss the problem, since they all presuppose the invariance of what we call today 'perception', as if it were not strongly influenced and even shaped by the historical emergence of deterministic mechanism. And, precisely for that reason, this cannot be considered a merely psychological issue. Beyond these basic explanations of determinism depending on an organic, or more specifically anthropomorphic process of the reduction of complexity, an ideological explanation is needed which again returns to the particular structure and

---

[33] In the end, even this sociological hypothesis proves insufficient: 'If the socio-psychological conditioning of thought can explain the vicissitudes of the hylomorphic scheme, it cannot explain its permanence and its universality in reflexion' (ILFI 52). In fact, the problem can be solved only at a deeper level: the level of the physical analysis of the 'process of formation', i.e. of individuation.

[34] The essay is part of the 'rarefied' bibliography of Individuation (see note 13). The note to which De Broglie refers can be found in Bergson (1934), p. 61.

[35] Going through the different stages of development of causality, Piaget eventually clarifies his opposition to Kant's conception of cause as an entirely a priori form (Piaget 1927, p. 323). For Piaget 'The evolution of the notion of cause presents for children a characterisation very close to the one we examined in the notion of real' (Piaget 1927, p. 291). See also Piaget (1926).

history of early-modern science, where deterministic causality progressively became the very postulate for a kind of knowledge which has gained undisputed worldwide supremacy and shapes now the 'spontaneous philosophy of scientists'. This topic raises a wide range of issues, which one cannot challenge without conjointly referring both to the epistemological structure and to the historical conditions of the emergence of early-modern science. From this perspective, the emergence of determinism calls not only for biological, but also for a conjoint epistemological and historico-sociological explanation, which should also refer to the acceleration of the process of mechanisation and mathematisation of the world picture which started with the emergence of early-modern science.[36] In effect, as reality became increasingly transparent to the lens of deterministic mechanism, the epistemological 'split' between subject and object became more and more violent, and was projected to the heights of metaphysics thus originating the—empty and abstract—subject of science. In view of such a long term project I shall just advance here a working hypothesis which one can draw from Simondon's philosophy of individuation and from his main sources.

Despite Simondon's link with Merleau-Ponty, according to the former there is no possible overcoming of the alternative between determinism and indeterminism within the phenomenological horizon, while, on the contrary, phenomenology is still nourished by that alternative. The hypothesis could be advanced that in *Individuation* Simondon extended Merleau-Ponty's biological paradigm to matter itself, following the path opened by the Greek verb '*phyo*', 'which alludes to the vegetative' (Merleau-Ponty 1956–1960, p. 3). But this approach would hide the true divergence between the two thinkers, which one could resume as follows: Simondon's fidelity towards Merleau-Ponty's project cannot evade a radical criticism of 'the subjectivity implicit in all conceptions of the individual, physical or biological, in the current doctrines' (ILFI 321). Against any phenomenological epistemology, which in the end proves dualist, as far as it presupposes the horizon of conscience (however conceived, as lived experience, raw perception or even 'flesh'),[37] Simondon's philosophy of individuation entails a non-phenomenological 'split' between subject and object, which should be read against the backdrop of a different kind of separation. He collocates the split not between the finite conscience

[36] On this intriguing topic one should consider at least the classical Koyré (1948) and Dijksterhuis (1986). Some Marxist debates concerning the historical connection between the emergence of mechanicism and of capitalism, bourgeoisie and state (and therefore the relation between mechanistic science and bourgeois ideology) should be taken into account too (see for instance Freudenthal-McLaughlin 2009, Schiera 1978). Furthermore, I believe French epistemology has important things to say, since it strongly took into account the influence of techniques in shaping the epistemology of modern sciences since its beginning in seventeenth century.

[37] Simondon's answers to Berger at the *Société* are adamantine. Berger: 'I would like to pose the question. Where are you collocating conscience? Is it to be presupposed since the beginning?' Simondon: '[…] I think the dualism which opposes subject and object cannot be maintained, on the contrary it must be considered the result of a process of formation which, in this case, is the process of individuation. The word *ontogenesis* summarizes the question' (Simondon 1960, pp. 764–765).

and its infinite distance from the real, but rather between different levels of reality *and* knowledge. Simondon's critique of both determinism and substantialism depends in effect on a critical attitude towards the Cartesian heritage from which he came—as it happened for many scholars of his generation: i.e. from the teaching of Georges Canguilhem.[38]

On the topic of metaphysical dualism, Canguilhem's thought is particularly effective, as far as it shifts the discourse out of the split between knowledge/reality which feeds that kind of metaphysics: Canguilhem's concern is rather with the shift between different milieus. Human beings in fact refer to different levels of reality which depend on correspondent levels of knowledge and practises. It is as if living-man lived in-between different milieus without being able to make a synthesis of them. Here it is what Canguilhem points to: 'a general theory of the milieu of the human being as technician and scientist [*l'homme technicien et savant*] [...] remains to be elaborated' (Canguilhem 1952, p. 96).

In *Le vivant et son milieu* Canguilhem enacts a philosophical move starting from a concept he derived from the life sciences: the concept of 'milieu' (Canguilhem 1952, pp. 129–154).[39] The vital activity of invention is conceived by Canguilhem as the relational activity between the 'internal milieu' and the 'external milieu' of an organism, in which organs, technical instruments and, furthermore, symbols, languages, moral norms, laws etc. emerge, which mediate and transform the relationship between the individual and his milieu, thus shaping the very terms of the relation themselves. In this sense science is a further instrument for the resolution of vital problems. How is the objective reality of science born, then? Three levels are implied in the operation he describes borrowing from the German ethologist J. Von Uexküll (1934) the terms *Umwelt*, *Umgebung* and *Welt*.[40] The *Umwelt* is the result of cutting off a subjective milieu (a '*prélèvement électif*') from the 'geographical environment' operated by each living being; the *Umgebung* is the construction of the

[38] Georges Canguilhem (1904–1995) became professor of philosophy at Sorbonne in the 1950s, succeeding Gaston Bachelard as the director of the *Institut d'histoire des sciences*. A former hero of French antifascism, after the second World War he was a key figure in the French educational system and he inspired plenty of philosophers educated in the period (See Foucault 1978). By reinforcing Bachelard's 'materialist element', Canguilhem probably played a key role both in Althusser's 'rupture' with Bachelard (Balibar 1978, pp. 212–213) and in Simondon's elaboration of the philosophy of individuation.

[39] In fact, as Canguilhem explained, the term had mechanical origins which affected its reception in biology through Claude Bernard's conception of the organism's '*milieu interieur*' (see in particular *L'expérimentation en biologie animale*, in Canguilhem 1952, pp. 17–39).

[40] Jakob von Uexküll, a German biologist, made the concept of *Umwelt* a tool for ethologists and philosophers; Heidegger, for instance, used the concept to highlight the difference between human beings and animals which 'do not have a World'. His little book *Streifzüge durch die Umwelten von Tieren und Menschen,* was translated in French, according to the Heideggerian inflection, as *Mondes animaux et monde humain* (1934). Not only does the original title not suggest any difference between human beings and other animals, but it also uses the term 'subject' when referring to organisms, whether *homo sapiens* or not. Von Uexküll was a key author for more than a generation of French philosophers from Canguilhem to Deleuze (e.g. they both derive from him the well-known example of the milieu of the flea), and also including Merleau-Ponty and Simondon.

'geographical environment' typical of human species, i.e. human *Umwelt*; the *Welt* is the 'universe of science' which is in fact the particular milieu of the human being 'as a scientist' (Canguilhem 1952, pp. 144–145, 152–153). In short, we must conclude that through science human beings built a 'universe of phenomena', with the aim of explaining the relation between the geographical milieu and the subjective milieus through natural laws. That is how through physics some human groups have produced 'objective reality' as an instrument for the solution of vital problems.

The peculiarity of physical science as an instrument, is that it has allowed humans to include the shift between subjective milieus (*Umwelt*) and a geographical milieu (*Umgebung*) in a *Welt* which can—in line with principles—be adopted by the whole species to augment its predictive capacities. Hence the universalising power of science, 'the birth, becoming, and progress [of which] must be understood as a sort of adventurous enterprise of life' (Canguilhem 1952, p. 153). Although deeply rooted in life, the functional determinism of physics tends therefore to 'dissolve living beings, which are centres of organization, adaptation, and invention, into the anonymity of the mechanical, physical, and chemical environment' (Canguilhem 1952, p. 153), thus reducing the relationship between the internal and external milieu to a mechanistic one, and subsequently hiding the very origin of this picture in collective invention. Therefore an epistemological problem is presented by mechanistic science, which, precisely because of its proved efficacy, tends to assume a metaphysical attitude.

On the contrary, according to Canguilhem, life sciences cannot but recognise themselves as the activity of an organism and are therefore compelled to pose the problem of this structuring shift between the living function of its look and the objective world in which it is placed. The problem biology poses to philosophy is the one concerning the ontological status of the human species milieu (*Umgebung*). Which is the relation between *Umgebung*, the *Umwelt* of the single subject-organism and the objective *Welt* of science? The emergence of these questions is for Canguilhem the mark of the peculiar reflexive characterisation of life sciences, after the 'milieu of the human being as technician and scientist' is born. The objective *Welt* of sciences is an epistemologically problematic reality of which only life sciences clearly show the reflexive nature, by mirroring the impossible reduction of life both to the objectivity of *Welt* (which is, in fact, their effect) and to the subjectivity of *Umwelt* (which is, in fact, their material condition). In short, what we call reality, which we know and live in, would be in this sense the intertwining of different milieus, none of which can claim ontological primacy. This unavoidable internal tension shifts the concept of 'geographical milieu'—against the mechanistic origin of the *term* milieu —to the original meaning of the *concept*, conceived in its relational meaning as what is '*between-two centres* [*entre-deux centres*]' (Canguilhem 1952, p. 130).[41]

[41] 'The individuality of the living [...] is not a being or a thing in the way which things were conceived in the eighteenth century. It always appears as a simple 'term in relation', with the other term being constituted by the 'milieu'. At each level of the living, one would discover such a relation that is constitutive of its proper terms. This is why 'the internal milieu of the organism' cannot be identified to the 'exterior' physical milieu, as Claude Bernard first proposed' (Lecourt 2012, p. 178).

In short, it is by scrutinising the deterministic 'model' from the viewpoint of the human species' relation to milieus, that one can test the limitations of the model itself. It is an instrument, ultimately grounded on human technical capacity in relation to nature, from which (techno) science stemmed as a—constitutively risky—operation of knowledge. Consequently, it cannot exhaust the complexity of the form of life it depends on:

> If science is the work of a humanity rooted in life before being enlightened by knowledge, if science is a fact in the world at the same time as it is a vision of the world [...] the milieu proper to human being is not situated within the universal milieu as contents in a container. A centre does not resolve into its environment. A living being is not reducible to a crossroads of influences. (Canguilhem 1952, p. 154)

In a patently phenomenological perspective Kojève had posed analogous questions on determinism in *L'idée du déterminisme* (1932), by differentiating the 'macroscopic experience (biological)' and the scientific one (Kojève 1932, p. 308). There he opposed the 'global given [*donné global*]'—i.e. the 'biological given [*donné biologique*]' of the 'concrete human being as a psycho-physiological being'—to the 'geometric data (mathematical) which one can isolate from [it]'. Being a mix of the two, he concluded, the objective 'universal given [*donné universel*]' of physics correspond to what a 'subject type' perceives, and which 'might—in principle—be extended to any "normal" human being' (Kojève 1932, p. 320). Kojève patently inscribed his argument in the phenomenological horizon, as an openly admitted 'metaphysical interpretation':

> In physical reality there is a qualitative difference of elements, there is interaction, effective action. Physical reality forms therefore a specific ontic region [of the 'ideal experimenter'], which is—so to speak—placed between the region of the mathematical [of the 'geometrical subject, symbolically represented by a system of coordinates'] and that of the biological [the 'concrete human (type)']. (Kojève 1932, p. 321)[42]

It is precisely Canguilhem's philosophy which, in a certain sense, permits the transposing of this hypothesis in a materialistic framework, thus thoroughly immersing science in vital processes and social dynamics. Canguilhem remained faithful to his original intent, even when the role he attributed to life-sciences was over, due to the recent developments of molecular biology, which seemed to entirely include the phenomenon of life in the deterministic milieu of physics. In effect, when Canguilhem's ancient project could no longer 'be taken up by the notion of individuality', he shifted to the theory of information and the concept of code.[43] Nevertheless, if 'the specificity of vitalism' is—as Lecourt poses it—the intellectual

---

[42] Kojève in fact refutes the label of 'idealism', and he prefers 'ideal-realism', underlining that physical reality is 'the result of a real interaction between the observing system (the physical subject as a physical reality) and the observed system', in which 'the frontier between the two systems can be moved, but never suppressed' (Kojève 1932, pp. 322–323).

[43] For a synthetic account of Simondon's use of the concept of code in relation to Canguilhem and to the theme of determinism, even if seemingly restricted to human individuation, see Morizot (2012), pp. 29–31.

demand to resist the 'subordination to a philosophy of Being', i.e. 'a substantialistic ontology', then the misadventures of Canguilhem's vitalism should be taken seriously, since they might lead 'to the elaboration of a new—non-Aristotelian—notion of form' (Lecourt 2012, pp. 182–183).[44] This is precisely what Simondon tried to do in his *Individuation in Light of the Notions of Form and Information.*

If Canguilhem proceeded to the analysis of the 'milieu of the scientist' on the basis of his philosophical analysis of the organism, Simondon in a way extended this very philosophical operation to 'all the domains of being', through the concepts of form and information, thanks to the reference to quantum physics. In effect Simondon deepened Canguilhem's criticism of mechanicism, by shifting his questioning of the subject-organism towards matter itself. Beyond the epistemological reflexivity Canguilhem attributed to biology, Simondon opened the possibility of extending to physical sciences the same 'encounter between history and its object' (Macherey 1998, p. 179), insofar as he considered the emergence of systems and their unfinished reconfiguration through processes of individuation as the basis for both ontological and epistemological questions: 'ontogenesis precedes critique and ontology' (ILFI 284). As a result, in Simondon's philosophy all sciences share the same reflexive, and therefore problematic, nature.

Thus the anti-substantialistic and anti-deterministic epistemology Simondon presents in *Individuation* seems to play a demystifying and anti-ideological function. It programmatically dismantles the 'conceptual couples' (form/matter, active/passive, subject/object, liberty/necessity) which have for centuries grounded a whole series of false alternatives, either opposing the individual to its milieu or dissolving it into the latter. And first of all, *Individuation* contributes to the overcoming of the conceptual and political machine represented by the ontological opposition between liberty and necessity, the key institution of a supposed ontological difference between the human being and nature. He meant to dismantle this dualistic 'machine' neither by reducing human beings to any imagined natural determinism nor by saving their alleged metaphysical nature. According to Simondon, 'neither totality nor indivisibility are possible in the case of human beings. An ontology, in the case of human being, would be an anthropology. Now, I do not believe an anthropology is possible; this is a postulate' (Simondon 1960, p. 756).[45]

---

44 A similar view on the exigency carried on by vitalism is expressed by Prigogine and Stengers' interpretation of Diderot's 'dream': 'Diderot's vitalist protest against physics and the universal laws of motion thus stems from his rejection of any form of spiritualist dualism. Nature must be described in such a way that the human being's very existence becomes understandable. Otherwise, and this is what happens in the mechanistic world view, the scientific description of nature will have its counterpart in the human as an automaton endowed with a soul and thereby alien to nature' (Prigogine-Stengers 1985, p. 83).

45 Simondon's 'difficult humanism' is crucial to the analysis of Barthélémy (2008), pp. 151–157. On Simondon's philosophy of 'human individuation', see Guchet (2012). On the basis of Combes (1999), Stengers proposes 'not (simply) a humanism "without humans", but another ontology, another anthropology, with other obligations' (Stengers 2004, p. 62). I try to show (Bardin 2013) how Simondon's humanism develops into a 'difficult ontology' which might prove compatible with materialism today.

This entails a hybrid net of socialised organisms producing, at different levels, milieus made of both reality *and* knowledge, as the outcome of different modes of existence of collectively shaped subjects the articulated emergence of which Simondon tried to describe through the notion of the 'transindividual'.[46] From his perspective, the emergence of a 'subject of science' within the modern dualistic framework appears to go far beyond the exigency of a technical relationship between human beings and their milieu, and rather appears the absolute, 'all encompassing' pure counterpart of nature as a deterministically conceived object. In fact, both constitute the non existing limit-cases of a relation in which social and natural processes dangerously play their productive role out of any effective control, precisely because of the fixed dualistic image that ideology projects on them, hiding their actual becoming.[47]

On the contrary, Simondon's philosophy of individuation struggles to maintain determinism in its technical, inventive and entirely epistemological dimension, out of any metaphysical assumption of ontological determinism. In a similarly accurate fashion, from a dossier dated February 1945, Canguilhem attacked

> The kind of Laplacean determinism [which] entails a conception of the relationship between human being and reality analogous to the Newtonian conception of the relationship between God and Universe [...] Human intelligence is understood as an *imitation*, and as a *limitation* of divine intelligence. The observers and measurers of the universe are exterior to the universe. (Canguilhem 1945, p. 9)

And yet Canguilhem continued to denounce the rhetorical tool of anti-scientistic indeterminism, whereby the rigour of scientific knowledge would be lost, thus opening a path to the myth of political voluntarism. In this sense it is worth recalling how—with as much exactitude and severity—in the same dossier he schematically noted the political costs of deriving indeterministic ontological conclusions from a conjoint critique of substantialism and determinism:

> Exploitation by fascism of some possible interpretations of the new discoveries in physics [...]—dissolution of the concept of *individuality*. Individuality destroyed at the ultra

---

[46] On the contribution that Simondon's analysis of the notion of the transindividual might add to this discourse by simultaneously conveying the issues of biology, technicity and the social exchange of significations, it is enough to say that the process through which a living being actually *becomes* a subject—i.e. the 'transindividual individuation' of a subject—is shaped by a paradoxical topology, since the process itself 'is not exterior to the individual and nevertheless it is partially detached from it' (ILFI 281). The first elaboration of Simondon's concept of transindividual is credited to Balibar (1993) and Stiegler (1994) and it was later developed by Combes (1999). For an inscription of the concept in the larger horizon of Simondon's epistemological and political thought, see (forthcoming) Bardin (2015).

[47] This conception of the emergence of modern dualism is perhaps compatible with what Latour (1993) named the 'modern constitution'. According to Latour modern science is the expression of a particular form of life productive of its own objects, in relation to which reality remains structurally at distance. As he explains, the theoretical assumption of the opposition between the human being and nature has gone together with the practical negation of it, with the subsequent proliferation of hybrids out of any political control.

microscopical level […]—liberty in the object itself. Therefore two arguments: against individualism → liberalism; against materialism → Marxism. (Canguilhem 1945, p. 10)

From Canguilhem's theoretical stance we can derive that determinism has to be programmatically adopted, although with the caution required since when, from the seventeenth century onward, the conceptual instruments of modern science have massively participated in social ontogenesis, with direct efficacy not only at the level of technological development, but also at the level of ideology.[48] From this perspective, the subject of science can be interpreted as the ideological projection to the heights of theory of the epistemological shift between the point of view of the individual-organism and the absolute point of view required by deterministic knowledge. A point of view which is, on the contrary, collective and always in the process of individuation. In this sense dualistic metaphysics, born from the reflection on modern science by some of those who invented and practised it, can be interpreted as the attempt to stabilise this shift by hypostatising it, thus making of the mathematisable reality of primary qualities—i.e. the 'milieu of the scientist'—reality in itself. To this reality, a deterministic *Res extensa* provided an adequate ontological repair, while to its scientific knowledge *Res cogitans* offered, on the other hand, a metaphysical umbrella which saved the subject of science from the vicissitudes of collectivity and of history, in which it is nevertheless embodied.

As one must admit, Thom had sensed who was responsible for this dangerous 'deviation of epistemology': 'one must think of a guilty party: is it Bachelard, with his benevolent smile […] ?' (Thom 1990a, pp. 77–78). In a different sense, with the resolution of those who intend to solve the risky enigmas their masters posed, in his philosophy of individuation Simondon was in search of 'a non Cartesian epistemology, borrowing Bachelard's expression, neither conceived in the sense of determinism nor in the sense of indeterminism' (ILFI 144), which could preserve the efficacy of science as a weapon against the *ideology of determinism* without defending the *counter-ideology of indeterminism.*

## References

Althusser, L.: Philosophie et philosophie spontanée des savants, 1967. Maspero, Paris (1974)
Amsterdamski, S. et al.: La Querelle du déterminisme: philosophie de la science d'aujourd'hui. Gallimard, Paris (1990)
Amsterdamski, S.: Halte aux espoirs, silence aux accusations. In: S. Amsterdamski et al. (eds.), pp. 228–246 (1990)
Bachelard, G.: L'Activité rationaliste de la physique contemporaine. PUF, Paris (1951)
Balibar, É.: From Bachelard to Althusser: the concept of "epistemological break". Econ. Soc. **7**(3), 207–237 (1978)

[48] As an exemplary case, one should think of the striking importance determinism had in the shaping of the new 'civil science' born with Hobbes' attempt to import Descartes' dualistic mechanism into political theory.

Balibar, É.: 'Spinoza: from Individuality to Transindividuality.' Mededelingen vanwege het Spinozahuis 71, Delft, 1997 (conference held on 15th May 1993)
Bardin, A.: De l'homme à la matière: pour une "ontologie difficile". Marx avec Simondon. Cahiers Simondon **5**, 25–45 (2013)
Bardin, A.: Epistemology and Political Philosophy in Gilbert Simondon: Individuation, Technics, Social Systems. Springer. (1st ed. Epistemologia e politica in Gilbert Simondon. Individuazione, tecnica e sistemi sociali. FuoriRegistro, 2010) (2015)
Barthélémy, J.-H.: Simondon ou l'encyclopédisme génétique. L'Harmattan, Paris (2008)
Bontems, V.: Echelles d'individuation, complexité et morphogenèse. Cahiers Simondon **2**, 81–97 (2010)
Bouquiaux, L.: L'harmonie et le chaos: Le rationalisme Leibnizien et La 'nouvelle science'. Peeters, Louvain-Paris (1994)
Bergson, H.: La pensée et le mouvant. PUF, Paris (1934)
Canguilhem, G.: Le normal et le pathologique, p. 1966. PUF, Paris (1943)
Canguilhem, G.: Dossier 'Déterminisme et indéterminisme' (février 1945). In: Archives de Georges Canguilhem G.C 11.3.7. ENS Paris, CAPHES (1945)
Canguilhem, G.: La connaissance de la vie. Paris: Vrin (1952)
Combes, M.: Gilbert Simondon and the Philosophy of the Transindividual. MIT Press, 2012 (1st ed. Simondon. Individu et collectivité, pour une philosophie du transindividuel. Paris: PUF) (1999)
De Broglie, L.: Physique et Microphysique. Albin Michel, Paris (1947)
Dijksterhuis, E.J.: The Mechanization of the World Picture. Pythagoras to Newton. Oxford University Press, Oxford (1986) (1st ed. Mechanisering van het wereldbeeld. Meulenhoff, Amsterdam)
Foucault, M.: 'Préface à l'édition américaine de Le Normal et le Pathologique.' In: Dits et écrits 1954–1988, vol. III. Paris: Gallimard. 1994, pp. 429–442 (1st ed. 'Introduction' to G. Canguilhem, On the Normal and the Pathological. Reidel, Dordrecht)
Freudenthal, G., McLaughlin, P.: The social and economic roots of the scientific revolution. Texts by Boris Hessen and Henryk Grossmann. Springer, Berlin (2009)
Garber, D.: Descartes' metaphysical physics. University of Chicago Press, Chicago (1992)
Garber, D.: Descartes Embodied: Reading Cartesian Philosophy through Cartesian Science. Cambridge University Press, Cambridge (2001)
Goldstein, K.: The Organism: A Holistic Approach to Biology Derived from Pathological Data in Man. Zone Books, 1995 (1st ed. Der Aufbau des Organismus. Den Haag, Nijhoff) (1934)
Guchet, X.: Merleau-Ponty, Simondon et le problème d'une "axiomatique des sciences humaines". L'exemple de l'histoire et de la sociologie. Chiasmi Int. **3**, 103–129 (2001)
Guchet, X.: Technology, sociology, humanism: Simondon and the problem of the human sciences. SubStance **41/3**, 76–92 (2012) Project MUSE
Kojève, A.: L'idée du déterminisme dans la physique classique et dans la physique moderne, p. 1990. Librairie Générale Française, Paris (1932)
Koyré, A.: 'Du monde de l' « à-peu-près » à l'univers de la précision'. In: Etudes d'histoire de la pensée philosophique. Armand Colin, Paris, 1961 (1st ed. Critique 28) (1948)
Laplace, P.-S.: Essai philosophique sur les probabilités. Paris: Bourgois, 1986 (5ème éd) (1814)
Latour, B.: We Have Never Been Modern. Harvester Wheatsheaf, New York (1993)
Lecourt, D.: The question of the individual in Georges Canguilhem and Gilbert Simondon. In: De Boever, A. et al. (ed.) Gilbert Simondon. Being and technology. Edinburgh University Press, pp. 176–184 (2012)
Lewin, K.: Principles of Topological Psychology. McGraw-Hill, New York (1935)
Macherey, P.: Georges Canguilhem's philosophy of science: epistemology and history of science. In a Materialist Way. Selected Essays by Pierre Macherey, pp. 161–187. Verso, London (1998)
Macherey, P.: Althusser and the concept of the spontaneous philosophy of scientists. Parrhesia **6**, 14–27 (2009)
Merleau-Ponty, M.: La Nature. Notes. Cours du Collège de France. (éd D. Sèglard) Paris: Seuil, 1995 (1956–1960)

Morizot, B.: Le *hasard contraint* comme modalité de l'individuation. Cahiers Simondon **4**, 9–31 (2012)
Petitot, J.: Identité et catastrophes (Topologie de la différence). In C. Lévi-Strauss. L'identité. Paris: PUF **1983**, 109–156 (1975)
Petitot, J.: 'Elogio della modernità. Il reincanto tecnico-scientifico del mondo nel pensiero di Gilbert Simondon.' In: Minazzi F (ed.) Realismo, illuminismo ed ermeneutica: percorsi della ricerca filosofica attuale. FrancoAngeli, Milano, pp. 89–107 (2004)
Piaget, J.: La représentation du monde chez l'enfant. F. Alcan, Paris (1926)
Piaget, J.: La causalité physique chez l'enfant. F. Alcan, Paris (1927)
Prigogine, I., Stengers, I.: Order out of chaos. Man's new dialogue with nature. Fontana, London (1985)
Prigogine, I.: 'La querelle du déterminisme, six ans après.' In S. Amsterdamski et al. (1990) cit., pp. 247–265 (1990)
Schiera, P. (ed.): Manifattura, società borghese e ideologia. Savelli, Roma (1978)
Simondon, G.: (ILFI) L'individuation à la lumière des notions de forme et d'information, Millon, Grenoble 2005. (1st ed. PhD thesis, 1957. In Archives de Georges Canguilhem. Paris ENS, CAPHES: GC. 40.2.1) The volume also contains 'Histoire de la notion d'individu' (ILFI 339–502), 'Note complémentaire sur les conséquences de la notion d'individuation' (ILFI 503–527), 'Forme, Information, Potentiels', a paper delivered in 1960 at the Société Française de Philosophie (ILFI 531–551), 'Analyse des critères de l'individualité' (ILFI 553–558), 'Allagmatique' (ILFI 559–566)
Simondon, G.: 'La psychologie moderne' (with F. Le Terrier). In: Daumas, M. (ed.) Encyclopédie de la Pléiade—Histoire de la science, pp. 1668–1703. Gallimard, Paris (1957)
Simondon, G.: Du mode d'existence des objets techniques. Aubier, Paris (1958)
Simondon. G.: 'Débat' following 'Forme, information et potentiels' (published in ILFI without the débat), pdf online version: http://www.sofrphilo.fr/telecharger.php?id=82: 751–765 (1st ed. Bulletin de la Société française de Philosophie 54/1960: 174–188 (1960))
Simondon. G.: 'Résumé de la séance de travail sur L'amplification dans les processus d'information.' In Le concept d'information dans la science contemporaine, 5ème colloque philosophique de Royaumont. Cahiers de Royaumount. Philosophie 5. Paris: Minuit, p. 417 (1962)
Simondon. G.: 'Culture et technique.' Bulletin de l'institut de philosophie morale et enseignement. Université libre de Bruxelles 55-56/1965, 3–16 (1965)
Simondon, G.: 'Initiation à la psychologie moderne. Première partie.' Bulletin de Psychologie 5/254 1966: 288–298; 8-9/256 1967: 479–490; 23-24/262 1967: 1449–1459 (1966–1967)
Simondon, G.: L'individuation psychique et collective. À la lumière des notions de forme, information, potentiel et métastabilité. Aubier, Paris (1989)
Simondon, G.: Communication et information. Cours et conférences. Chatou: La Transparence. (2010)
Stengers, I.: Comment hériter de Simondon? In: Roux, J. (ed.) gilbert simondon: une pensée opérative, pp. 300–315. Publications de l'Université de Saint-Etienne, Saint-Etienne (2002)
Stengers, I.: Résister à Simondon? Multitiudes **18**, 55–62 (2004)
Stiegler, B.: Technics and Time 1: The Fault of Epimetheus. Stanford University Press, 1998 (1st ed. La technique et le temps 1. La faute d'Épiméthée. Paris : Galilée) (1994)
Thom, R.: 'Préface' to Laplace (1814) cit (1986)
Thom, R.: 'Halte au hasard, silence au bruit.' In S. Amsterdamski et al. (1990) cit., pp. 61–78 (1990a)
Thom, R.: 'Postface au débat sur le déterminisme.' In S. Amsterdamski et al. (1990) cit., pp. 266–279 (1990b)
Thom, R.: 'Morphologie et individuation.' In Gilles Châtelet et al. (ed.) Gilbert Simondon. Une pensée de l'individuation et de la technique. Paris: Albin Michel, pp. 100–112 (1994)
von Uexküll, J.J.: Streifzüge durch die Umwelten von Tieren und Menschen. Springer, Berlin (1934) (French transl. Mondes animaux et monde humain, Suivi de La théorie de la signification. Denoël, Paris, 1965)

# Chapter 2
# How to Invent a Form: An Inquiry into Gilbert Simondon's Philosophy of Perception

**Giovanni Carrozzini**

**Abstract** In this article I provide a historical-critical inquiry into Simondon's work about the invention and the production of perceptive forms provided with informational potential, thus underlining the effects of undetermined factors during the process of perception. According to Simondon, during the operation of perception, perceptive images/forms are invented especially thanks to elements that cannot be reduced to a principle of order and simplicity. From this perspective, perceptive images or perceptive forms (these two terms are used by Simondon in a synonymic way) are not only the simple product of the application of innate and determined schemes: they are, on the contrary, an inventive result including "undetermined" elements. This can be especially observed in the case of the long duration perception of a turning linear object, as Simondon notes in his article *La perception de la longue durée*(1969–1970). Before analysing Simondon's peculiar viewpoint about perception, I analyse Simondonian theory of changes, namely allagmatic, to show his original position and his distance from determinism and pure indeterminism.

## 2.1 Introduction

In this article I provide a historical-critical inquiry into Gilbert Simondon's work about the invention and production of perceptive forms provided with informational potential, thus underlining the effects of *undetermined* factors during the process of perception. Before analysing Simondon's peculiar viewpoint about perception, I offer a short outlook of the main aspects of Simondon's philosophy, in order to show his original position and his distance from determinism and pure indeterminism.[1]

[1] It is important to underline that Gilbert Simondon does not offer any precise definition of determinism and pure indeterminism in his works, so they could be considered as synonymous of causalism and casualism. Following Simondon, in this article I follow this *undefined* use of the terms, without providing any further definition of determinism and pure indeterminism.

G. Carrozzini (✉)
Università del Salento, Lecce, Italy
e-mail: giovannicarrozzini@gmail.com



A. Sarti et al. (eds.), *Morphogenesis and Individuation*,
Lecture Notes in Morphogenesis, DOI 10.1007/978-3-319-05101-7_2

In the *first section*, I analyse Simondonian concept of *allagmatic*, namely the *theory of changes* of a structure in an operation and vice versa. This theory is conceived by Simondon to describe the internal processes which take place during an operation of individuation. The purpose of this first part is to underline the differences between this peculiar philosophical viewpoint and determinism and pure indeterminism.

In the *second section*, I provide an inquiry into Gilbert Simondon's viewpoint about perception, conceived as an operation of restraint individuation, namely an individuation which takes place within an already individuated domain, i.e. the biological one. I sum up Simondonian criticisms to associationism and *Gestalt-theorie* and I underline his original proposal to modify the concept of *pregnant perceptive/images forms*.

In the *third section*, I offer an example of *concrete perceptive situation*, starting from Simondonian analyses of *long duration perception*, focused on *undetermined factors* which interact during this process.

## 2.2 The Distance Between Simondon's Allagmatic and Determinism and Pure Indeterminism

> Determinism and indeterminism are just borderline cases; because there is a development of systems: this development is the one of their individuation: there is a reactivity of the systems in relation to themselves. The evolution of a system would be determined only if there was not an internal resonance of the system, that is to say no exchange between the different levels which are hold within it and by which it is composed; in this case, no quantum exchange would be possible and you could know the development of this system thanks to a theory of the continuous, or thanks to the law of large numbers, such as in the case of Thermodynamic theory. Pure indeterminism would coincide to a huge internal resonance, thanks to which every modification arriving on a determined level would immediately reverberate on every level disguised as a change of structure.[2]

Simondon's philosophy is focused on internal processes of "operations of individuation"; we can conceive his peculiar viewpoint as an endogenous perspective fixed on all kinds of physical, biological, psychical, collective and technical procedures which produce, after their own achievement, the structuring of a new field of reality. As I shall show in the *second section* of this article, the analysis of "forms in themselves" is not the real purpose of Simondon's inquiry, unless they still contain potentials to trigger new individuations; otherwise, they are considered by Simondon just as "vestiges" of achieved individuations. His attention, as I said, is focused instead on processes and operations of individuation. When he analyses, for example, the *process offormation* (*prise de forme*)—that is a peculiar kind of individuation, i.e. the one of artificial realities, such as produced objects—taking the case of a brick of clay, he concentrates his own attention *within* this process, criticizing Aristotle's hylomorphic theory. He notes that the hylomorphic

---

[2] Simondon (2005), pp. 148–149.

philosophy supposes form and matter conceived as the two principles of individuation, instead of considering the internal dynamics of this operation of individuation. Such an operation is, on the contrary, the relation between these two scales of reality and the inner state of the system in which they interact. Form and matter cannot explain the *process* and its real endogenous dynamics, as he remarks in the *First chapter* of the *First part* of his *Thèse d'Etat*, *L'individuation à la lumière des notions de forme et d'information*[3]: form and matter are just theoretic schemes, aiming at representing (and simplifying) real composite half-chains, while individuation is the communication between these two scales of reality—i.e. the molecular (matter) and the molar (form) ones—and the state of the system in which they interact and communicate one another.

In other words, individuation can be conceived as the structuring process of a field of reality, which develops *within* a system. This system must be in a metastable state of equilibrium; metastability is a state of equilibrium full of potentials and virtualities. Individuation starts from a local (but not defined) trigger, which is called *singularity*, the activity of which spreads in a progressive way. Simondon names this progression transductive propagation, which spreads *within* the system, producing a new state of equilibrium, that can be, after this former individuation, the place of other operations of individuations only if it keeps on providing within it some unachieved potentials and virtualities. This is not just a simple process of *formation* (judged from an external point viewpoint), as the philosophical tradition uses to define this procedure, basing on Aristotelic and Thomistic concepts of form, matter, individual substance (*synolon*) and *materia signata*. On the contrary, this is a process of *in-formation*, which can develop and arrive to an achievement by means of the presence of potentials and virtualities, which are the internal conditions and proprieties of the system in which an individuation can be produced.

According to Simondon, determinism and pure indeterminism are just two ways to analyse and to explain processes, such as individuation, from an external viewpoint: Aristotelic theory is a deterministic, causal way to explain and to provide knowledge of a phenomenon, such as individuation, which, on the contrary, cannot be known and analysed with traditional theoretical categories. As Simondon writes in his *Thèse d'Etat*:

> we cannot *know*, in the traditional meaning of this word, *individuation*; we can just individuate, individuate ourselves and individuate within ourselves; this conception is on the border of knowledge as we use to consider it; it is an analogy between two sorts of operations; it is a certain way of communication.[4]

According to Simondon, this is the reason why determinism (i.e. all sorts of causalism) and pure indeterminism (i.e. casualism) cannot be considered as correct approaches to the inner dynamics of individuation: they are, in fact, traditional ways for *grounding knowledge*, i.e. to make phenomenon easily understandable and intelligible to human faculties. According to Maurice Merleau-Ponty's

[3] Simondon (2005), pp. 39–66.

[4] Simondon (2005), p. 36.

phenomenology, conceived as a "direct description of [the] experience,"[5] we can say that, from Simondon's viewpoint, also individuation—conceived as an operation—can just be *described*, but it cannot be really *known*, neither explained with the traditional theories of knowledge nor graphically represented. His philosophical work on individuation provides the "tools" for a description of this process, but it does not offer any manner to know it in a traditional way of thinking *knowledge*. Simondon calls his own theory *allagmatic*, which means *theory of changes*; it is fixed on the description of the change of a structure in an operation and of an operation in a structure: this is not a traditional theory of knowledge; it can be maybe conceived as a sort of *keeping-in-touch* with processes from an internal, endogenous viewpoint. Employing a Sartrian expression, individuation is a *situation* that cannot be explained: it can just *take place*.

According to Simondon, determinism and pure indeterminism start from the consideration of a form, trying to *explain* the genesis of this form or its own changes: Simondon maintains that forms are just partial results of processes, or *vestiges*, and he underlines the fact that the philosophy of individuation, i.e. *allagmatic*, must not consider forms as a starting point of its inquiries; it should instead put itself *within* the operations that, *at least*, can provide forms. As I said, when individuation takes place, the considered system must be in a metastable state of equilibrium, namely it must be far from stable or instable state of equilibrium: it must hold potentials and virtualities. Its inner state can also be named a "disparate" condition, that is to say that there is not a "phase coincidence" between the "portions" of this system. Without a phase difference the system cannot go beyond its state of saturation, thanks to the action of a singularity, and thus no individuation can be produced.

According to this peculiar perspective, determinism and pure indeterminism start from "establishing" a form, i.e. a structure, to explain the way in which it arises or the way in which it evolves or develops. For this reason these structures or forms can be represented in a more or less defined way and so they can be also translated in a graphic model or in a mathematical explanation of its procedure of change. On the contrary, Simondon states that there are some elements of the process which cannot be graphically represented: these elements are the ones that allow the trigger of the process, that is to say potentials, virtualities and spatial localisation of singularity activity. They are peculiar of the contingent "situation" of a considered system and of its specific state of equilibrium: they can be neither determined nor represented. However, this procedure may not even be explained by a purely indeterministic perspective, because of the presence of vaguely defined structures which are within the system or arriving into it (if they are external to the system) as in the case of the structure of a singularity. In a specific case, this singularity can be conceived as the defined structure, i.e. provided with defined proprieties, of a seed crystal introduced (or already hold) in a solvent. Individuation is a (inter)mediation, i.e. a relation between different orders or scales of reality provided with their own

[5] Merleau-Ponty (1945), p. 10.

structures and dynamics, which interact in a systemically way, thus producing a field of interactions and of changes of structures in operations and vice versa. As I said, Simondon's inquiry does not offer instruments to represent these processes: we can define allagmatic as a real non-representative theory, since it does not provide graphic, mathematical or logic schemes to *explain* and to represent operations of individuations. As he writes in the *Fist chapter* of his so-called *Third part* of his *Thèse d'Etat*: "potentials, precisely because they are potentials and therefore they are not structures, cannot be represented as graphic elements of the situation."[6]

According to Simondon, determinism and pure indeterminism consider at first a more or less defined structure as a model or a scheme of their own inquiries, trying to represent the genesis of a structure or its transformation even in a graphic way. But, as Simondon observes in a short text entitled *Allagmatique*, written in the same years of his *Thèse d'Etat,*[7] these approaches are classical ways of explanation adopted by the *science of structures*, which developed in the seventeenth century with Galileo and Descartes; he maintains that sciences should also develop an inquiry into operations; the synthesis of these two perspectives—so called *allagmatic* inquiry—cannot preserve classical methods. Determinism, pure indeterminism, deductive and inductive methods cannot be considered useful approaches to individuation, because they are theories of explanation and theories of knowledge, i.e. their own viewpoint in an external one, i.e. *a posteriori*. This is the reason why they are inadequate to *describe* individuation. In other words, Simondon—as Jean-Hugues Barthélémy notes in his work *Simondon ou l'encyclopédisme génétique*[8]—is a sort of pioneer of the philosophy of Complexity, in the sense that he discusses and rejects traditional analytic approaches to reality.

In the case of every considered peculiar situation of system and of their metastable states, complexity is due to the presence of potentials and virtualities that are not representative or determined factors. These potentials and virtualities are however the pre-conditions to trigger an individuation, which is a sort of in-formational chain reaction, the dynamics of which cannot be known. What you could know are just final forms: but after the achievement of an individuation, forms or structures are just a sort of discard, the stability of which is the evidence of an already started thanatological process, unless they still contain some other potential. In any case, form is not the central topic of the philosophy of individuation: moreover, Simondon's theoretical attention is not really focused on "changing forms as a way of explanation of these forms",[9] as René Thom interprets Simondon's inquiry, mainly because it is not concerned with *explanation* at all, but with

[6] Simondon (2005), p. 238. This chapter is the *First paragraph* of the *Second chapter* of the *Second part*, entitled *L'individuation des êtres vivants*, of the *Thèse*, according to the original typed version of Simondon's *Thèse d'Etat*, a copy of which is preserved in the *Archives Canguilhem*—Fond Caphès and another one in the *Fond Hyppolite* of the Ecole Normale Supérieure of Paris.

[7] Simondon (2005), pp. 559–566.

[8] Barthélémy (2008).

[9] Thom (1994).

*description*. The genetic operation of structuring forms, that is to say individuation, according to Simondon, is a description of a process *a praesenti*. According to Simondon, in fact, even the act of knowledge, namely the peculiar situation in which something can become intelligible, is an operation of individuation. *Within* an act of knowledge, conceived as an operation of individuation, you cannot distinguish the subject from the object (once more the *form* from the *matter*) because the subject and the object of knowledge are closely linked one another within this process. Only if one considers the process from above, reflecting and explaining the process, one can speak of subject and of object: this is the reason why Simondon's philosophy cannot be considered as a representative model of the concrete "situation" of knowledge *in progress*.

After having analysed the notion of *allagmatic* in Simondonian work, in the next section I shall focus on his philosophy of perception.

## 2.3 Form and Information in Concrete Perceptive Situations

Which are the theoretical consequences of Simondon's viewpoint in the specific case of philosophy of perception? In this article, I intended to assume this topic as an emblematic case in order to show the distance between determinism, pure indeterminism and Simondon's philosophy of individuation, concerning the process of invention of perceptive forms. It is necessary to explain what Simondon means with the concept of perception and what is really new in his perspective about this classical theme for philosophy. In the already quoted *Third part* of his *Thèse d'Etat*, Simondon holds that perception is a *functioning*—I should say, in a modern way of speaking, a *cognitive process or procedure*—that works within the psychical domain of living beings. It is important to remark that psychical individuation is not a main individuation: the only two kinds of main individuations you can find in Simondon's work are physical and biological individuations, conceiving the second one as a sort of *inchoative* physical individuation, namely an individuation that does not pursue its own achievement, its own stability. As he notes in his work: "the living domain would be a physical domain in suspension, slowed in its own process and indefinitely extended."[10] On the contrary, psychical individuation is a more restricted operation rather than physical and biological ones; it arises within the biological domain of living individuals, that is to say into an already individuated domain, when problems that are posed by the *milieu* to individuals cannot be solved by means of their biological dynamics. Psychical individuation is also named by Simondon "individualisation", i.e. an individuation triggered within an already individuated field of reality, i.e. the biological one. Moreover, psychical individualisation is not a specific or special prerogative of human beings; as Simondon

[10] Simondon (2005), p. 152.

explains in his work on individuation, all species of animals—in some peculiar and contingent cases—can develop psychical ways, i.e. psychism, to solve the external problems posed by the *milieu* in which they are sited. From this specific viewpoint, the differences between the animals, conceived as living beings which can trigger psychical individualisations, are just differences of degree: they are not qualitative or hierarchical differences. Perception is maybe the psychical functioning to which Simondon accords more attention with respect to the other psychical operations. But how does he analyse perception? Namely, which are the aspects on which does he focus is own interest? Even in this peculiar case, what he is interested in is the process of genesis of forms, more than to forms on their own as a result of processes. As he states in the *First paragraph*, of the chapter *L'individuation des unités perceptives et la signification*, he aims at offering a description of the process of "segregation" of perceptive units, that we can also call *perceptive images/forms*—according to Simondon's idea that you cannot perceive without imagining, even if (as in the case of sensory-motor images) you can imagine without perceiving.[11] His inquiry is focused on the genesis of coherent perceptive images/forms, cached and described *within the process of their own production* and not at the end of this process. This process—he states—is a perceptive *situation* and it should be described *in progress*, that is to say in a concrete perceptive context. That is why Simondon states that most of the theories of perception do not consider living beings and—in this peculiar case—human beings in their own real perceptive situations, but just in a scientific and analytic context, such as the one of laboratory.

At the beginning of this paragraph, Simondon seems to sum up and develop the criticisms that Maurice Merleau-Ponty moved against associationism in his *Phénoménologie de la perception*; but Simondon's viewpoint goes further; he also criticizes *Gestaltpsychologie* and its laws of perception. According to Simondon, the greatest limit of associationism lays in his inability to offer an account of inner coherence of perceptive units; as he maintains in this fist paragraph of this chapter, associationism "does not explain why the individualised object has an inner coherence, a substantial link that can give it an effective interiority that cannot be considered as the result of the previous association."[12] This kind of criticism was already moved to this psychological theory from Maurice Merleau-Ponty, in the *Second paragraph* of the *First chapter* of his work of 1945.[13] Simondon and Merleau-Ponty criticize the idea that perceptive units coherence, according to associationism, would be a sort of indirect consequence of *habit* and of *memory*; associationism is in fact an application to the domain of scientific psychology of the traditional theories of modern empiricist philosophy, such as the thought of John Locke and the one of David Hume. Empiricism found in similarity, comparison and communion the laws that explain the process of so-called complex ideas and in the habit the ground for causal explanations. These theories, applied to the domain of

[11] See Simondon (2008).

[12] Simondon (2005), p. 233.

[13] Merleau-Ponty (1945), pp. 20–33.

scientific psychology, maintain that perceptive units are the results of the association of sensible data. According to Simondon, these theories do not explain why these units have "real" inner coherence, that is to say why they show an internal organisation and a stable configuration. Maurice Merleau-Ponty opposed to this hypothesis *Gestaltpsychologie* viewpoint: he remarked that perceptive field shows a coherent organisation because of the systemic organisation of the units that we can perceive. Simondon goes further, criticizing also *Gestalttheorie* and especially its law of *Prägnanz*. As I said at the beginning of this article, Simondon develops a sort of endogenous perspective that aims at analysing processes and operations *a praesenti*, namely *in the middle of their development*. According to Simondon the psychologists of *Gestaltpsychologie*, especially at the beginning of their theoretical work, did not offer an adequate description of genesis of forms, even if they offered to the domain of scientific psychology two important notions, such as *isomorphism* between the inner operations of physical fields and the functioning of all other kinds of natural processes (i.e. physical and biological ones) and as the notion of *psychical field* (derived from Thermodynamic theory). He notes that *Gestalttheorie* should be considered as a sort of *deterministic* way to account for perceptive process and for structuring perceptive image/forms configurations. According to *Gestalttheorie*, perceptive units are *cached in* the external *milieu* and this phenomenon derives from a sort of deal between internal mental schemes and the external configurations and organisations of the elements forming a system. This is the main reason why the most *pregnant* forms—that is to say the most *significant* ones—are the *simplest*, the *most stable* and the *most organised* forms, i.e. forms which stand out among all other possible configurations in a perceptive process. Simondon observes that this peculiar viewpoint does not consider some factors: these factors interact in the operation and in the concrete situation of perception, that is to say *within* the system formed by perceiving *subject*, i.e. psychical individual in the act of perceiving, perceptible factors organized in configurations in external world, and the relation of these *two half-chains* of this process. Perception and, at last, segregation of perceptive units, conceived as perceived images/forms, are not just the result of a deal between determined or determinable external schemes of configurations and internal mental schemes: it is on the contrary a phenomenon that provides some other factors, which can be defined *relational undetermined factors*, which specify concrete and contingent perceptive situations. As I said, these relational factors could be named, according to Simondon, *undetermined factors*, even if his peculiar viewpoint, also in the case of perception, cannot be conceived as a pure indeterministic way to describe the process, because of the presence of defined structures interacting into the operation. These remarks about the presence of *undetermined factors* in a concrete perceptive situation just serve to Simondon's purpose to underline the inefficacity and the insufficiency of purely deterministic perspectives to describe concrete perceptive situations. As he writes in his *Thèse d'Etat*:

> the absolute set, in the physical world, is not just formed by the solved body and the solving body: it is the solving, the solved and the set of forces and potential energies that are translated by the word metastability, applied to the state of metastable over-saturated situation at the time in which the beginning of crystallisation operates. At this time of

> metastability no deterministic "Good *Gestalt*" is adequate to form what is produced; the phenomenon such as epitaxy show the existence at a critical moment […] of a sort of relative uncertainty of the result.[14]

According to Simondon, neither determinism nor pure indeterminism can describe the *work in progress* of a concrete perceptive situation; determinism *explains* results of perception as the emergence of a well-defined, stable form, cached in the world, the proprieties of which would pre-exist to the act of perception, namely within the inner organization of the set of perceptive configurations, the schemes of which *fit with* mental ones; pure indeterminism would *explain* perception as a process provided with casual dynamics that cannot be described. Even in the domain of perception, allagmatic provides a description of perceptive context, conceived as a system of interactions between structures and operations, which transform into one another. In other words, in deterministic and in pure indeterministic ways of providing an *explanation* of real perceptive contexts, a *dark zone* lies: this dark zone is the process on its own, that is to say *the individua(lisa) tionof perceptive units*:

> [i]f form was really given and pre-determined, no genesis would be possible, no plasticity, no incertitude about the future of a physical system, of an organism, of a perceptive field […]. There is a genesis of forms as there is a genesis of life.[15]

Simondon's viewpoint is focused within this *critical moment* which is the *work in progress* of this genesis. From this peculiar perspective concerning the question of perceptive process of genesis of image/forms, *pre-determined* forms, namely stable, pre-organised, and *shaped forms* cannot be considered as real forms: they are just "structural schemes and not forms. It is possible—as Simondon remarks—that these structural schemes are inborn; but they are not enough to explain segregation of units in perception."[16] According to Simondon, perception is not just an activity of caching forms from outside world: perception is an operation of individua(lisa) tion of images/forms. These images/forms cannot pre-exist to the act of perception in a concrete perceptive situation; that is why they are *invented* in the process of perception: "perception is not caching a form, but it is the solution of a conflict, the discovery of compatibility, the *invention* of a form."[17]

Even if perception is a restraint kind of individuation, however it shares with authentic individuations the prerogative to trigger a new equilibrium, a new state of system (i.e. physical and biological ones), in order to produce, to *invent* a new form, the genetic processes of which can be considered neither determined nor representable in a traditional way of explanation, i.e. in a mathematical or in any graphic way.

---

[14] Simondon (2005), p. 234.

[15] Simondon (2005), p. 233.

[16] Simondon (2005), p. 236.

[17] Simondon (2005), p. 235.

This is the starting point of his criticism against one of the essential laws of perception according to *Gestaltpsychologie*, i.e., as I said, the so-called law of *Prägnanz*, that is also known as *law of Good* Gestalt. This law is one of the main laws of perception, analyzed by the *Gestalt* psychologists since the beginning of the twentieth century: a "Good *Gestalt*" is a defined form, the pregnancy of which derives from its simple, coherent, regular and determined organization. In other words, *Gestalt* psychologists maintain that living beings organize their perceptive images according to standardized patterns—seemingly innate—such as geometrical forms (circle, triangle, square, etc.). According to *Gestaltpsychologie*, perception seems to be based on a biological deterministic law of order and simplicity, which reduces the complexity of perceptive contexts: the most pregnant form is the simplest one. Moreover *Gestaltpsychologie* focuses his inquiries on the perceptive interaction between living beings and "their own worlds", assuming an "exo-genetic" viewpoint to describe perceptive experiences. As Paul Guillaume, the psychologist who disseminate *Gestalttheorie* in France at the beginning of the twentieth century, notes in 1937, in his work *La Psychologie de la Forme*: "psychical facts are *forms*, that is to say *organic units that individualize themselves in the spatial and temporal field of perception and representation.*"[18] Even if Gilbert Simondon shares with Guillaume and the psychologists of *Gestalttheorie* the assumption that perceptive units can be considered as organisms (provided with a sort of autonomy and inner organization and configuration), he rejects the idea that their genesis would be just the emergence of the most stable and determined form above the others. "Best forms—as Guillaume observes—are the *regular, simple, symmetric ones*. The perceived form is the best one (law of good *Gestalt*) [...] all elements [...] are more efficient when they are linked with symmetry, less efficient if they are in conflict with it."[19] According to *Gestaltpsychologie*, the rules of proximity and similarity of the elements into a determined configuration (in a shaped organization) make their grouping pregnant, i.e. significant and meaningful. Simondon asks a question about this strictly determined way to analyze perceptive processes: "are circle or square, conceived as examples of "good *Gestalten*", that is to say "pregnant", stable, organized, simple forms—superior to the forms that an artist invents?",[20] cached in the act of their production, namely *during their genetic process*. By asking this question, Simondon wants to underline the fact that the law of "Good *Gestalt*" seems to be based on the idea that forms mentally preexist and predetermine the operation of perception: that is why the resulting perceptive world seems to be a strictly predetermined and regular one, in which there is no place for singular and undetermined factors. On the contrary, he states that some undetermined factors interact during the process of perception. As he writes in his work:

[18] Guillaume (1963), p. 22. This citation is taken from the Italian translation of Guillaume's work of 1937.

[19] Guillaume (1963), pp. 58–59. This citation is taken from the Italian translation of Guillaume's work of 1937.

[20] Simondon (2005), p. 236.

"the square, the circle, and all the simple and pregnant forms are just structural schemes, rather than forms."[21] According to Simondon, during the operation of perception, perceptive images/forms are invented especially thanks to factors that cannot be reduced to a principle of order and simplicity. Orientation of the factors in the set, polarity, tensions, duration of perception can be considered examples of undetermined factors which interact *within* the process of *individuation* of perceptive images/forms. That is why forms invented by artists can be considered *more provided with information* than *well-shaped* geometrical forms, like a square or a circle. In short, perceptive images/forms are considered by Simondon as inventive solutions to problems, and the process of perception is considered an *in*-formational procedure. In a concrete perceptive situation, living beings do not know what they have to perceive, such as—on the contrary—in the most of the experiments in laboratory, where the subject of observation is more or less informed about what he has to do or to say. And even if instructions are not clear or explicit, he finds himself in an artificial context; he knows, for example, that he is observed and that his behavior is detected and measured; according to Simondon, "lots of experiments in laboratory, which take a subject that is not so much tense, perfectly assured does not accomplish the conditions in which genesis of forms takes place."[22] The law of Good *Gestalt* considers just the cases in which the subject knows what he has to do, not only in an artificial context, such as the one of laboratory, but in all the situations in which instructions are clear; real conflicts between living beings and their environment—such as his perceivable context—cannot be produced in predeterminate contexts, because they are not *real perceptive situations*. Real conflicts cannot be produced into these contexts and I would say that, according to Simondon, when there is not a conflict you cannot have a real perceptive process. Perceptive processes—conceived as solutions of conflicts and as inventions of a new order of reality, i.e. operations of restraint individuation—are solutions to conflicts: they need a condition of "psychical tension", or, as Simondon writes: "a degree of metastability. […] This is the reason why the laws of good *Gestalt* are not adequate to explain the segregation of the units in the perceptive fields; they do not consider in fact the meaning of the solution given to a problem offered by perception."[23] The forms invented by an artist can be conceived, according to Simondon, as real forms, because they are inventive solutions to perceptive conflicts, derived from potentials and vitualities, which are inner conditions of a metastable system. In other words, they are not just structural pre-determined or geometrical schemes, such as a circle, a triangle or a square: real pregnant forms are not best forms because they are stable and symmetric configurations of elements; they are solutions provided with meaning for a specific, concrete conflict and this meaning cannot be reduced to geometrical and deterministic patterns. On the contrary, this meaning is the *partially undetermined result* of a perceptive system of psychical

[21] Simondon (2005), p. 237.
[22] Simondon (2005), p. 235.
[23] Simondon (2005), p. 235.

tensions, conceived as the degree of metastability. Forms as circle, triangle and square are not invented forms in perception, for they do not retain the *uncertainty* of the previous potentials and virtualities of the system from which they arise. *Gestalttheorie*, according to Simondon's viewpoint, focuses on "transformation and degradation of forms rather than on their genesis."[24] This remark underlines the distance between Simondon's philosophy and all deterministic (and *pure* indeterministic) approaches to forms, which start from forms to explain their genesis, but also from those approaches that consider just their transformations in a fully representative way. Invented forms are, on the contrary, polarizations of creative activity, which are not representable in a traditional way, cached in its *critical moment of production* and *within their own genetic process*, providing a solution to a contingent problematic situation that needs a psychical effort, and not just the commitment of living forces or biological tendencies. These invented forms are partial results of an in-formational process, i.e. of a perceptive situation. But invented forms, that is to say the most "pregnant ones", are not the most stable ones: they are perceptive images/forms that maintain a great level of information for the perceiving subject concerned in a perceptive systemic situation. These information give and assure meaning to the solution but they are not even transferable to other situations: "law of *Prägnanz*", in the new meaning that Simondon accords to it, is strictly linked with the undeterminable factors of systemic contexts. The degree of metastability is not a definable state, such as stable equilibrium, and that is why Simondon seems to reject the possibility to represent, for example in a mathematical way, the real endogenous dynamics that work within a so-configured process. What science can *explain* are just the "vestiges of the individuation already achieved in a metastable state."[25] Simplified forms, such as stable forms, are schemes with very low information content: on the contrary, invented forms, conceived as real pregnant and meaningful forms *in progress*, are provided with a huge degree of information. Their origins should be found in the undetermined factors interacting during the process of perception, such as potentials and vitualities, but also as *duration of perceptive situation*, which are *non-representable* factors.

Summing up, the perceptive images, or perceptive *forms* are not only the simple product of the application of innate and determined schemes: they are, on the contrary, an inventive result including "undetermined" elements. Simondon tried to reproduce these concrete perceptive situations, making an experiment in a very unusual way to observe the processes working in the progress of a *long duration perception* of a turning linear object or of a turning simple geometrical figure. In the *third section* of this paper, I shall sum up the contents of a Simondon's scientific article, *La perception de la longue durée*, focused on the attempt to reproduce a *concrete perceptive situation* in a laboratorial context.

[24] Simondon (2005), p. 235.

[25] Simondon (2005), p. 237.

## 2.4 Long Duration Perception: An Example of Concrete Perceptive Situation

In his scientific article *La perception de la longue durée*, reporting the effects that can be produced in a laboratorial perceptive "situation", Simondon—taking himself as the subject of his own scientific observations—notes that *long duration perception* can modify the perception of an object, interfering in this process as a real *undetermined* factor of system. In this peculiar laboratorial context, the duration of perceptive process in not pre-determined at the beginning of the experiment: "[during] a continuous observation, experimentally prolonged for many minutes or even for many hours, it can arise some phenomenon which do not arise during tachistoscopic arrangements or [during] middle durations."[26] Simondon aims at showing how a strictly scientific method should be implemented by some factor which exceeds this perspective. This unusual kind of experiment, which opposed Simondon's method of observation to the one that was used by most of the psychologists of his university (Sorbonne-Paris V), such as Paul Fraisse, is based on the idea that extending the duration of the presentation of a perceivable *stimulus* in a laboratorial context—without employing tachistoscopic methods, such as in the case of most of *gestaltic* and behaviorist scientific experiments—it was possible to recreate a tense perceptive situation. This situation must be a metastable system in which undetermined factors could provide to the invention of meaningful forms, conceived as solutions of perceptive problems. As Simondon writes in his article, during a long duration perception, "you can observe a genesis of the object, at first as an individualized unit and then as an organized unit, provided with a structure and a movement."[27] Moreover:

> perceptive-motor relation between a living being and his *milieu* retains, in an original and universal way, some less instantaneous detection and identification processes, because they are less institutionalized and less culturalized or conventional. These processes, which aim at the detection and identification of living world, arise during long duration perception […]. Because of subsequent saturations, after having exhausted the meaning of a conventional and artificial stimulus conceived as a symbol or as a sign, the perceptive activity individualize itself in a unique structure, and then it concretizes itself in an organism.[28]

Long duration perceptions try to recreate in an artificial context the situation of tense system in which subjects loose their awareness of a *safe condition* and self-consciousness of the instructions and stimuli. In this peculiar case, it is possible to *describe* a real genesis of invented perceptive images/forms, because of the interaction of undetermined factors which provide the invention of pregnant and meaningful solutions, namely organic perceptive units provided with information, without representing the result, but just describing some working effects within the

[26] Simondon (1969), p. 368.

[27] Simondon (1969), p. 398.

[28] Simondon (1969), pp. 399–400.

process. It is true that in this specific case, these observations seem to be a report or an *explanationa poteriori* of a phenomenon working *a praesenti*. But the undefined conditions of experimental set, such as the no pre-determined time duration of observation and the fact that the subject of the experiment is at the same time the observer, acting the role of undetermined factors in the process of inventing forms, do not provide representative structural schemes, graphically representable, but just the description of proceeding dynamical effects.

Also on the basis of what has been learned from this experiment, the law of *Prägnanz* in the Simondonian definition seems to require a new criterion to evaluate invented perceptive images/forms: this criterion can be neither quantitative nor qualitative, as in the case of traditional methods of associationism and *gestaltic* theory. According to Simondon, this criterion, that must preserve the same undetermined feature of concrete perceptive situations and of their metastable state of systems, is the one of *intensity of information*. As Simondon writes in his *Thèse d'Etat*: "a very contrasted photograph, that is to say with a very violent light and shade effect, or an out of focus one can have more value and intensity than the same photograph in a perfect graduation, i.e. respecting the value of every detail or the geometrically centre of the photograph without deformities."[29]

[29] Simondon (2005), p. 242.

Invented perceptive images/forms, conceived as real pregnant forms—because of their indeterminate degree of intensity of information—can be conceived as *real concrete individualized forms* (i.e. contingent solutions to perceptive concrete situations) *just for the time of their genesis*, that is to say *within the working process of their individualisation* or if they are able to preserve, within their configuration, some potential factors of their previous and genetic context of production. The process once achieved, you can find, in some specific cases, just structural schemes, but not real perceptive images/forms, conceived as informational systemic processes in progress.

## 2.5 Concluding Remarks

According to Simondon, determinism and indeterminism just consider the end or the beginning of a process, as in the case of all kinds of explanations and reflections, but *living forms*, conceived as *living processes*, such as a perceptive operation of individualisation, are *in the middle*. They are, maybe, describable, but not really understandable: this is the reason why Simondon's philosophy of individuation cannot be conceived as a traditional theory of knowledge, even in the case of his analysis of perceptive processes. Simondon tries to put his own viewpoint *within* the working processes to *describe* the *genesis* of forms. This genesis, conceived as an operation of individuation, cannot be known (in a traditional way), because of the presence, *within* the process, of *undetermined, non-representable* factors.

## References

Barthélémy J.-H.: Simondon ou l'encyclopédisme génétique. L'Harmattan, Paris (2008)
Guillaume, P.: La psicologia della forma, it. tr. of Giulio Gnoli. Editrice universitaria, Florence (1963)
Merleau-Ponty, M.: Phénoménologie de la perception, Gallimard, Paris (1945)
Simondon, G.: La Perception de la Longue Durée. J. de Psychol. Normale et Pathol. **4**: 397–419 (1969)
Simondon, G.: L'individuation à la lumière des notions de forme et d'information, préface de J. Garelli. Millon, Paris (2005)
Simondon, G.: Imagination et invention (1965–1966), edition établie par Nathalie Simondonet présentée par Jean-Yves Chateau, Chatou, Editions La Transparence (2008)
Thom, R.: Morphologie et individuation. In: Châtelet, G. et al. (eds.) Gilbert Simondon. Une pensée de l'individuation et de la technique. Paris, Albin Michel, pp. 100–114 (1994)

# Chapter 3
# Individuation and Semiogenesis: An Interplay Between Geometric Harmonics and Structural Morphodynamics

**Alessandro Sarti and David Piotrowski**

**Abstract** The process of individuation deals with the genesis and the constitution of forms and how they differs from each other. Gilbert Simondon in his 1958 thesis L'individuation à la lumière des notions de forme et d'information (Simondon 2005) shows how individuation has a pivotal role in various phenomenon at a physical, biological, psychophysical and social level. The concept of form is often used in a synonymous manner to structure, instead, here it is utilized with a very different meaning. Structure refers to a codified, pre-existing form while individuated form is a continuously emerging reality.

## 3.1 Individuation and Semiogenesis

### *3.1.1 Individuation of Forms and Spaces*

The process of individuation deals with the genesis and the constitution of forms and how they differs from each other. Gilbert Simondon in his 1958 thesis L'individuation à la lumière des notions de forme et d'information (Simondon 2005) shows how individuation has a pivotal role in various phenomenon at a physical, biological, psychophysical and social level. The concept of form is often used in a synonymous manner to structure, instead, here it is utilized with a very different meaning. Structure refers to a codified, pre-existing form while individuated form is a continuously emerging reality. For example G. Deleuze describes the difference between the musical forms of classical music (structured) and the

A. Sarti (✉)
Center for Mathematics CAMS, CNRS-EHESS, Paris, France
e-mail: alessandro.sarti@ehess.fr

D. Piotrowski
LIAS, CNRS-EHESS, Paris, France
e-mail: david.piotrowski@ehess.fr

 
A. Sarti et al. (eds.), *Morphogenesis and Individuation*,
Lecture Notes in Morphogenesis, DOI 10.1007/978-3-319-05101-7_3

individuated forms of the music of Pierre Boulez music, where time is not pulsed and scales are replaced by continuous variation of musical material (Deleuze 1978).

The epistemic framework of Simondon's thesis is that of quantum physics, which explain the formation of elementary particle starting from continuum operators. This is different to classical physics description, in which particles are given a priori. In quantum mechanics particles are individuated by spectral analysis of the underlying operator, that is the descriptor of the physical field. Particularly every measured configuration (particle) corresponds to a specific eigenstate of the continuum operator. In general Simondon is interested in the constitution of forms at biological, perceptual and social levels. In these situations morphological units are formed from relational operators that are defined as preindividual. Preindividual refers to an heterogeneous ensemble of relations that acts as a field of differences that pre-exist to individuation of the entities. Reality is operatorial because it produces the passage from a form to another by means of individuation processes derived from preindividual relations, overcoming the distinction between being and becoming. Deleuze and Guattari clarify the Simondonian concept of preindividual in Mille Plateaux (Deleuze 1980) and associate it with a dense relational set, the very heterogeneous and a-centered network they called Rhizome: "A Rhizome is made by plateaus", a concept taken from Gregory Bateson "to designate a continuum or intensity region, vibrating on itself and self-developing avoiding any orientation on a culminating point or external finality" (Deleuze 1980).

In Deleuze-Guattari the concept of the pre-individual is extended to discrete relational sets (networks), while the individuation process retains its harmonic characteristics, inherited from the epistemology of quantum mechanics. Namely individuated entities, defined as intensity or consistency planes, can be seen as the vibrational modes the relational network. Analogous to physical particles in quantum mechanics which are correspondingly defined as modes of vibration of the quantum mechanical operator. The harmonic aspect is always present throughout Simondon's work. In his 1958 thesis Simondon calls the holism-effect of form individuation "a resonance". On a physical level he addresses systems as existing at a metastable equilibrium point, in such a way that when the equilibrium is lost, solutions corresponding to the stable eigenfunctions tends to take over, individuating the consistent forms as vibrational modes of the system.

*In synthesis, we find a common perspective in Simondon, Deleuze-Guattari and Bateson, framing the individuation principle inside a relational epistemology. This allows us to think the pre-individual as a heterogeneous continuously changing relational field which is functionally supported by harmonic processes that act to individuate its consistent forms.* With this framework in mind we intend to reveal a functional/operatorial level of the individuation process in terms of its double articulation of (a) the definition of suitable relational graphs and (b) estimation of their reduced spectral structure by means of non-linear harmonic analysis.

Interestingly we will see that the process of geometric harmonic analysis leads to a reduced dimensionality representation of the relational graph, corresponding to John Duns Scotus idea of *contraction individuationis*, i.e. the idea that the individuation process creates a contraction (Scot 2005).

We will limit our self in this article to modeling perceptual individuation by outlining the generative role of the relational network, considered as a pre-individual stratum of relations. In Simondon's perspective, this involves not only morphological gestalt laws but also immanent and singular perceptive situations. For this reason the relational graph can be largely undetermined, dependant on the contingent situation of perception. Therefore the proposed model should be considered more as a theoretical scheme, rather than a positivist recipe for a deterministic solution. In addition, spectral individuation has a holistic character which produces the coherent organisation of the perceptual field and avoids the limitation of an associationistic approach, which was deeply criticized by Maurice Merleau-Ponty as well as by Gilbert Simondon (see for example the article of Giovanni Carrozzini in this volume). In the paper we would like to describe the two main conceptual objects that are generated by harmonic individuation: (a) forms and (b) reduced spaces, as well as the manner in which they are coupled to (c) spaces of control, seen as semiotic extension of reduced spaces. The first conceptual object refers to the constitutions of units as single eigenfunctions of the preindividual operator. Each eigenfunction is in direct relation with the formation of a perceptual unit in the gestaltic sense. The second conceptual object is the space (hyper-plane) generated by linear combination of the set of most important eigenfunctions. It is a space of positional information where every point is defined as a combination of eigenfunctions (an accord of resonant units) that intrinsically define the position of points in relation to the whole form. This set of eigenfunctions designates the principal axis for the semiotic formants, as seen in the case of the formemes of the visual background, in the tradition of the research group $\mu$ (Edeline et al. 1992).

Whereas the two first objects are intrinsic to the relational matrix, the third one extends the first two objects to a semiotic level by enlisting external dimensions: in fact, the reduced space is now equipped with appropriate potentials that allow for the space's partition, in this manner deploying a categorization process. External dimensions are the control variables of an internal dynamics space as in the tradition of the morphodynamical structuralism. This categorization process connects the essentially continuous nature of reduced spaces to discontinuous semiolinguistic forms. In other words partitioning of the control space sets the correspondence between gestalten and semiolinguistic units.

In the following section discuss the semiotic reasoning behind this functional extension, that breaks away from the logic of intrinsic individuation that is grounded on the relational matrix, thereby possibly appearing arbitrary or artificial.

## *3.1.2 The Semiotic Extension*

We can justify the aforementioned extension from a semiotic perspective in the following manner: (1) forms that are individuated by spectral analysis have a proto-semiotic nature; (2) a functional extension which preserves, prolongs and fulfills the (proto)semiotic nature of individuated gestalt into fully-semiotic units can be considered natural; (3) and the constitution of signs, particularly the discontinuous nature (we will see: the differential nature) of semiotic forms, can be obtained through a natural extension of spectral individuation. In order to consolidate this line of reasoning, we must first be more precise regarding the proto-semiotic nature of the gestalt level of spectral individuation with which the sign's structure must continuously prolong. And secondly, we must explain the need to introduce an upper level of functional organization in order to reach the sign level. First, therefore, we observe that the gestalt individuation displays a dual level of organization similar to that of semiotic entities. That is, at the pre-individual level, the relationships are carried by autonomous and unspecified units which, thereby, receive a relational identity. But, further, those relational units are relegated in a background for benefit of another object, an higher level object, for the realization of which they are a necessary condition, but which nevertheless exists as a whole whose characteristics cannot be found in any of the underlying relational units. Such an upper-level object, supported by the underlying relational units, can be seen as their meaning. This is the same as how Merleau Ponty describes the phenomenological characteristics of the structure form/background as the meaning of its underlying parts: "la couleur de la figure est plus dense et comme plus résistante que celle du fond; les bords de la tache blanche lui "appartiennent" et ne sont pas solidaires du fond pourtant contigu; la tache parait posée sur le fond et de l'interrompt pas, Chaque partie annonce plus qu'elle ne contient et cette perception élémentaire est donc déjà chargée d'un sens". See Merleau-Ponty (2012, p. 9). In addition, this unsymmetrical dual level of organization is similar to the phenomenological structuring of the linguistic sign, as seen in Husserls phenomenological analysis (Husserl 1969, 1991, 1993, 2000). Husserl indicates that when perceiving a sign, the material part of the sign (the concrete data) is effectively presented and perceived through its concrete qualities (sound or graphical material), and as such is also a condition of the realization of the sign. However, this concrete part exists only in a secondary manner, which is relative to an upper-level object towards which it directs the mind. Namely, the meaning, which consequently is located at the primary level of ones attention (a thematic level) in the awareness scope. Secondly, we have to justify the introduction of an external functional component to reach the sign level. One may observe that the meaning of a word (the signified) cannot be considered as an emerging value of a relational network whose units are concrete parts of words (word forms such as chains of phonemes, or, approximately, the signifier). It is obvious that pre-individuated units of the relational matrix, on one hand, and emerging spectral forms on the other, do not have the same relationships as concrete word forms and meaning values, simply due to the arbitrarily principle: word forms, as concrete data, have nothing to do with their meaning. Therefore, provided that relationships between words forms refer only to their concrete nature and properties, no meaning value will ever emerge from such a network. However, if we choose a structural perspective, the semantic values of words are given by their relationships or, in other words, by their position in the network they compose. In this case, indeed, we are no longer dealing with concrete word forms, but with true signs. Here, the direct application of the harmonic process fails. The reason is that, in this case, the semantic values of the words are directly embedded in the relationships they share in the network. And therefore, semantic values exist here independently of any harmonic process. What we learn here, is that the application of a harmonic approach to the semiotic field, based on gestalt proto-semiotic nature, needs to be extended in order to capture the signs level. Our question becomes: what kind of functional extension must be given to the spectral individuation process in order to achieve full semiotic structure? The major condition for this functional extension is to minimize arbitrarity, in other words, the functional extension must be close to the internal organization of the spectral object. This condition is particularly satisfied when part of the spectral object takes place in the semiotic system as one of its functional components. This component therefore constitutes a common term (an articulation) that unifies the spectral organization and the upper-level semiotic architecture.

In order to choose a semiotic architecture that satisfies this type of condition, we turn to the saussurian theory of sign and its morphodynamical characterization. We will see that the saussurian organization of the content plane (seen as a differential categorization that institutes semantic values of words: the signified) is governed by certain control parameters. We will also see that the axes where such control parameters are located are naturally indicated, in the spectral perspective, as eigenvectors. The connection between harmonic individuation and the saussurian theory of sign is then established as the identification of: (1) the consistency plane generated by the eigenvectors and (2) the support of a control space equipped with parameters that determine the dynamics at the source of the differential identities meanings (the signified). The saussurian theory of sign appears then to be a natural extension of the harmonic individuation processes. Consequently, also, the harmonic forms of proto-semiotic individuation appear to be naturally extended to differential (negative and oppositive) forms of individuation.

In the following section we would like to address some specific issues regarding perceptual individuation, meaning the constitution of forms in the domain between visual perception and signification. We describe perceptual individuation as an enlarged gestalt, representing not just morphological, but also broader, multimodal qualities. Appearing, from the beginning, as a union of perception, action and expression. Particularly, we focus on the constitution of figures in a gestalt sense, on the constitution of the formemes of the visual background, in the tradition of the group $\mu$ and on emergence of the articulation between figure and background as one of the first cognitive categorization processes.

## 3.2 The Individuation of Perceptual Units

In this section we will afford the issue of the constitution of visual units from the vast amount of local information that is present on the visual field. First we will return to a mathematical model of the functional architecture of the visual cortex, as introduced in Citti and Sarti (2006) and Sarti et al. (2008), constructing a suitable neurogeometrical structure. Secondly we will show how a visual stimulus excites the neurogeometry generating a relational field of linked elements, that constitutes our pre-individual field. Finally the vibrational modes of the relational field are considered as constituent of perceptual units in terms of principal components of the network. Even if in our analysis we only use morphological information, the relational field can be enriched by elements coming from other somatosensory sources and virtually implicate all the contextual elements of a situated and singular perceptual experience.

### *3.2.1 The Geometrical Structure of the Primary Visual Cortex*

The neuro-physiological findings show that for every point on the retinal plane there is an entire set of corresponding simple cells in the cortical layer that are sensitive to all the possible orientations of the stimulus (Hubel and Wiesel 1977). Many models of this structure have been presentented (Hoffman 1989; Petitot and Tondut 1999; Citti and Sarti 2006; Sarti et al 2008). In Citti and Sarti (2006) this structure has been well mathematized by the Lie group of rotation and translation E (2), that is a manifold associating to every point a position $(x, y)$ and a rotation $\theta$ (see Fig. 3.1).

Infinitesimal transformations connect to each other points of the group (simple cells) through the tangent planes generated by the left invariant fields $\vec{X}_1$ and $\vec{X}_2$

$$\vec{X}_1 = (\cos(\theta), \sin(\theta), 0) \quad \vec{X}_2 = (0, 0, 1). \tag{3.1}$$

Finally simple cells are locally connected by the integral curves of the field $\vec{X}_1$ and $\vec{X}_2$. The natural curves of the structure are the integral curves of the vector fields $\vec{X}_1$ and $\vec{X}_2$, starting at a fixed point $(x_0, y_0, \theta_0)$:

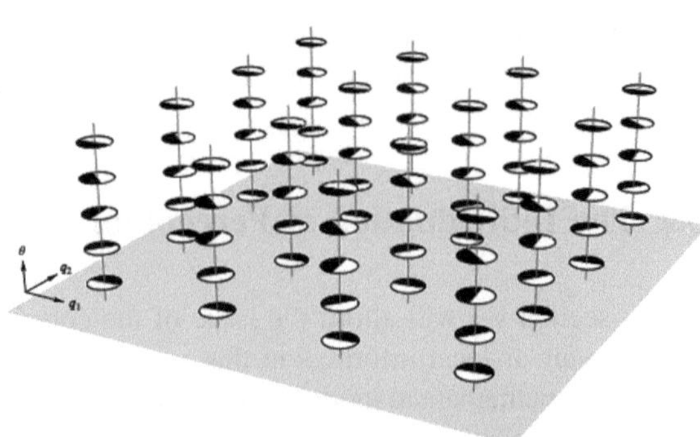

**Fig. 3.1** The Lie group E(2) modeling the set of simple cells receptive profiles in the primary visual cortex

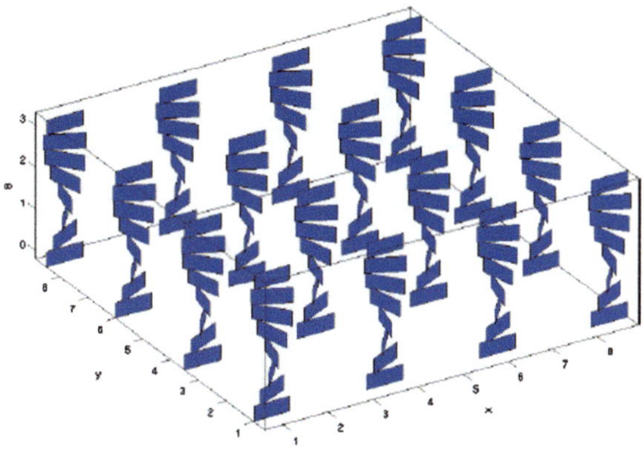

**Fig. 3.2** The tangent structure of the Lie group E(2) generated by the left invariant fields $\vec{X}_1$ and $\vec{X}_2$, modeling the infinitesimal connectivity between points of the group that correspond to simple cells

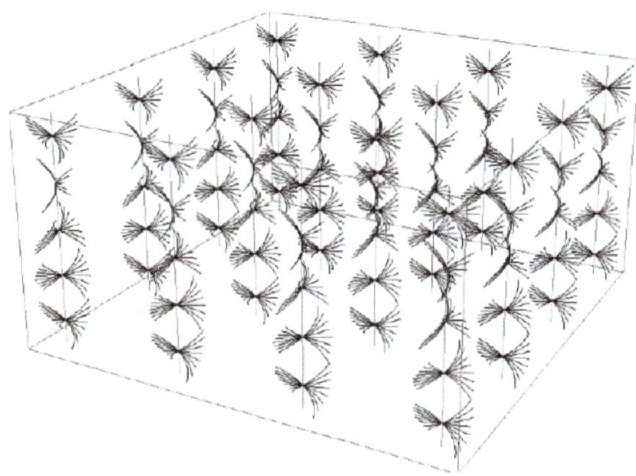

**Fig. 3.3** The integral curves of the left invariant fields $\vec{X}_1$ and $\vec{X}_2$, modeling the local connectivity between points of the group that identify simple cells. It defines the cortical connectivity graph

$$\gamma' = (x', y', \theta') = \vec{X}_1(x, y, \theta) + k\vec{X}_2(x, y, \theta) = (\cos(\theta), \sin(\theta), k) \qquad (3.2)$$

$$\gamma(0) = (x_0, y_0, \theta_0),$$

and obtained by varying the parameter $k$ in $\mathbb{R}$ (Fig. 3.4). These curves can be used to model the local association field as described in Field et al. (1993) (Figs. 3.2 and 3.3).

We can define the distance $\mathcal{D}_s$ between two points as the length of the shortest path that connects the two points. In the Euclidean case the minimum of the path length is obtained within all possible curves, while here we will minimize only over the set of integral curves of the vector fields $\vec{X}_1$ and $\vec{X}_2$. Using the standard definition we call length of any curve $\gamma$

$$\lambda(\gamma) = \int ||\gamma'(t)||dt = \int \sqrt{1 + k^2}dt. \qquad (3.3)$$

**Fig. 3.4** Given a random distribution of segments, certain perceptual units are perceived as coherent structures. For example a *semicircle* and a *vertical* segment (*left*). The image elements sample the cortical connectivity graph to obtain an immanent graph depending on the specific stimulus (*right*). The sampled elements are shown in *red*

It can be proved that the parameter $k$ expresses the curvature of the projection of the curve $\gamma$ on the plane $(x, y)$. A deeper analysis of variational calculus in this group structure can be found in Citti and Sarti (2006).

### 3.2.2 The Relational Field Induced by an Image

In the previous chapter we described the action of cells in response to a visual stimulus as the geometric lifting that takes place in the rotation-translation group. Every active cell corresponds to a point of the group, and their connectivity identify the neurogeometry in terms of integral curves of the sub-Riemannian Lie group structure. This structure can be represented as a graph whose nodes are the active cells $p_i = (x_i, y_i, \theta_i)$ and whose links are the neural connectivities with a weight $A_{ij}$ that is defined in terms of their distance

$$A_{ij} = e^{-\mathcal{D}_s^2(p_i,p_j)}$$

where $D_s$ is the sub-Riemannian distance defined above. (see Fig. 3.4). A similar graph can be obtained in terms of a probability density kernel in the case that a stochastic model of the connectivity issued. We refer the interested reader to Sanguinetti et al. (2010).

### 3.2.3 From the Relational Field to Perceptual Individuation

We would like now to investigate what is the role of the immanent relational field in the constitution of perceptual units. Human perceiving a scene, generally segment it easily into coherent segments or groups. There has been a tremendous amount of effort devoted to achieving the same level of performance in computer vision. The overall difficulty lies on the fact that perception is a global process as outlined by Gestalt theory. Visual perception acts as a differentiation process of the entire field of view, first articulating figure and background in a global way. Then, it proceeds for segmentation of single structures by subsequent differentiations.

During the chain of processing performed by the cell activity the unity of the original objects present in the image is completely lost, but at the end the object shows up again at the perceptual level as a unit. How is this reconstruction possible? This process is known as binding or perceptual grouping and it has been extensively studied at least from two different points of view: the experimental psychology of Gestalt, oriented to find the phenomenological laws of perceptual organization; and neuro-physiological studies that focused on biological functionalities underlying grouping. On a neurophysiological level there is a vast body of experimental evidence that indicates that the perception of a single object in a visual scene involves the response of a large number of neurons that are distributed over a large spatial region. A very accredited hypothesis is that visual morphologies are represented in the brain as an effect of the mean field of neural discharge either in terms of the synchronization (phase locking) of oscillatory neural responses Wilson and Cowan (1972), Wang (1995) or in terms of eigenfunctions of the neural mean fields equations (Bressloff and Cowan 2003). In Sarti and Citti (submitted) it has been shown that grouping can be achieved through the eigenfunctions of the previously introduced immanent graph $A_{ij}$:

$$\sum_j A_{ij} x_j = \lambda x_i \tag{3.4}$$

reframing the problem of constitution of perceptual units in terms of geometric harmonics, where the geometry is provided by the relational field. The segmentation of different objects corresponds to different eigenvectors $x_i$. In Fig. 3.5 the two principle eigenvectors of the immanent graph, represented by the matrix $A_{ij}$ are visualized in gray values, showing the global emergence of perceptual units.

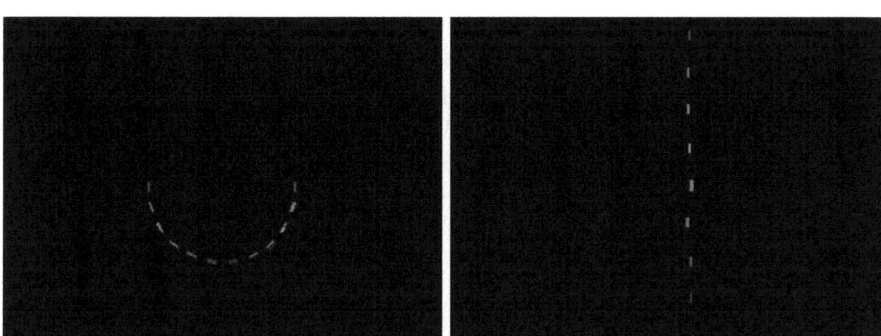

**Fig. 3.5** The spectral decomposition of the neurogeometric operator $A_{ij}$. The first eigenvectors correspond to the most salient perceptual units

We can note that this approach interprets the emergent figures in the image as eigenstates of the preindividual field. Mathematically this corresponds to the singular value decomposition of the previously defined neurogeometrical graph.

Models of image segmentation based on singular value decomposition, and in general on dimensionality reduction, are largely used in contemporary computer vision (see for example Neumann (1998), Shi and Malik (1997)), we also refer to Weiss (1999) for a review of methods.

## 3.3 The Formants of Visual Semiotics

In this section we will focus on the concept of visual forms as developed by the research group $\mu$, who have deeply studied the constitution of formants in visual semiotics, particularly in the *Traité du sign visuel* (Edeline et al. 1992). Even if the theoretical analysis of the group addresses the entire system of forms, colors and textures we will limit here just to the constitution of formants of plastic forms. We will briefly summarize certain elements of the group's analysis of formants and outline how they can be functionally constructed from a process of harmonic individuation acting on a relational field.

In the following the basic oppositional categories of the plastic semiotics is discussed.

### *3.3.1 First Opposition: Figure/Ground*

A fundamental problem the visual system solves is the segregation of objects from their supporting backgrounds. Both phenomenology and Gestalt psychology clearly recognized that the most fundamental kind of experience is that of the perception of a figure against a ground. Husserl interprets the problem from the point of view of environing things (Husserl et al. 1907):

"A perceived thing is never there alone by itself; instead, it stands before our eyes in the midst of determinate, intuited environing things".

Merlau-Ponty outline the quality of phenomenological experience when perceiving figures in terms of scale, shape and color constancy, while the background is instead perceived as indeterminate (Merleau-Ponty 2012). From the morphological point of view the Berliner Gestalt school has shown that perception tends to define that which is characterized by convexity, regularity, closure and symmetry, a 'figure'. A shape with these characteristics is called salient, because it pops-up from the indifferentiated background (Khler et al. 1929).

Let's look at how the figure/ground opposition is individuated from a suitable relational field. The boundaries of a solid figure, such as the black rectangle that is visualized in Fig. 3.6 (left) are easily detected by the action of odd simple cells, with the mechanism described in Sect. 3.2. The neurogeometry connecting the

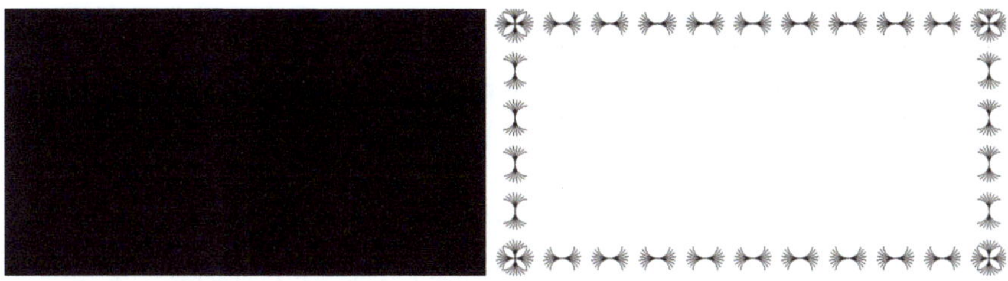

**Fig. 3.6** The relational structure (*right*) of the contour of a *rectangle* (*left*) as sampled from odd simple cells. No information is present regarding what is inside (figure) and outside (background) of the *rectangle*

activated cells is shown in Fig. 3.6 (right). If the spectral decomposition of this relational network is performed it becomes easy to check that the global boundary is individuated as a unified shape by the principal eigenvector. But let's noticed that this process well individuates the shape boundary without providing information about the inside/outside of the object. That is, the issue of figure/ground segregation is not still resolved.

To afford it let's consider the action of "even" simple cells, who retrieve the position of the boundary by considering also its polarity. As described in Sarti et al. (2008), maximally firing cells are positioned slightly offset (inside or outside) with respect to the object boundary. The resulting neuro-geometrical graph is constituted now by two layers of connectivity (represented in red and in blue in Fig. 3.7) related to odd simple cells with different polarity (Favali 2013). A spectral decomposition of the graph shows that the first two eigenvectors individuate separately the two connectivity layers and most importantly it will present them in the order: 1st eigenvector→figure, 2nd eigenvector→background.

This ordering resolves the figure/ground segregation problem, due to the fact that the interior relational graph is always stronger then the exterior one, as its elements closer to each other than the others. This individuation process takes into account all together the gestalt laws of good continuation, convexity and closure.

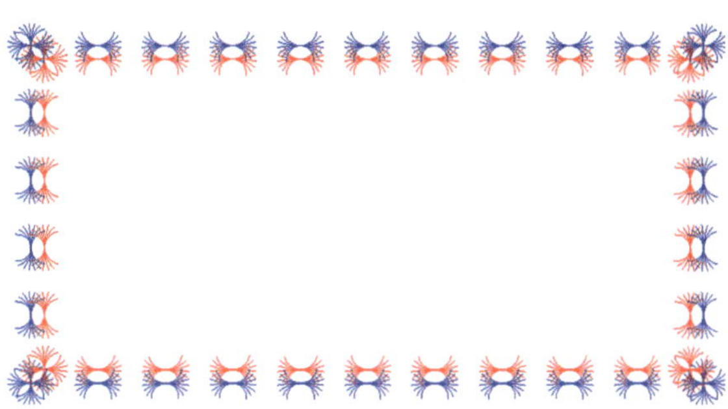

**Fig. 3.7** The relational structure of the contour of a *rectangle* as sampled from "even" simple cells. *Red* and *blue* colors denote the connectivity of "even" simple cells with different contrast polarity

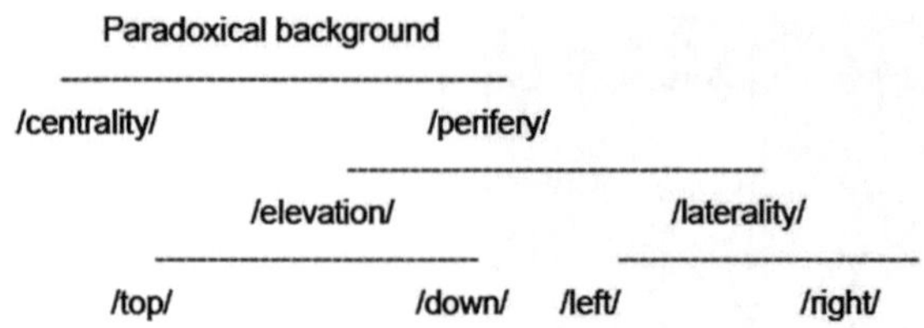


**Fig. 3.8** The structure is represented in Fig. 3.8 (from Edeline and Klinkenberg (1992))

### *3.3.2 2nd, 3th, 4th Oppositions: The Formemes of Plastic Forms*

In the analysis of formemes of the *Traité du sign visuel* the status of the background is very peculiar, since it is not considered undifferenciated and unlimited, or without defined boundaries, as for example in Rudolf Arnheim's *Toward a Psychology of Art* (Arnheim 1966). The goupe $\mu$ defines the background as a limited space, like a sheet of paper or a painting's frame. In this manner the background contains its own form and for this reason is defined as "paradoxical":

> The form of the background intervenes by imposing its laws on the other forms that detach from it Edeline et al. (1992).

The formemes indicate the relationship between the figure and the paradoxical background. Particularly the first formeme corresponds to the position of the figure in the background:

> It is clear that different significations emerge when the form is positioned in the center of a background or in a different point of the space (Edeline et al. 1992).

The structure of opposition of the formeme position is tripartite and it is articulated in the oppositions (Fig. 3.8):

- Second opposition: the figure could be central or marginal in relation to the center (centrality/perifery).
- Third opposition: if marginal, the figure could be above or under the center (verticality).
- Fourth opposition: if marginal, the figure could be to the left or to the right of the center (laterality)

In order to retrieve the basic oppositions of plastic semiotics in terms of individuation of forms, we will introduce a suitable relational field for the paradoxical background. The connectivity structure of the background is isotropic and delimited in space domain $\Omega$ in such a way that the connectivity weight between two points $p_i = (x_i, y_i)$ and $p_j = (x_j, y_j)$ is given by

**Fig. 3.9** An isotropic and delimited relational field defines the paradoxical background. A few connectivity kernels sampled from the surface are visualized in the figure

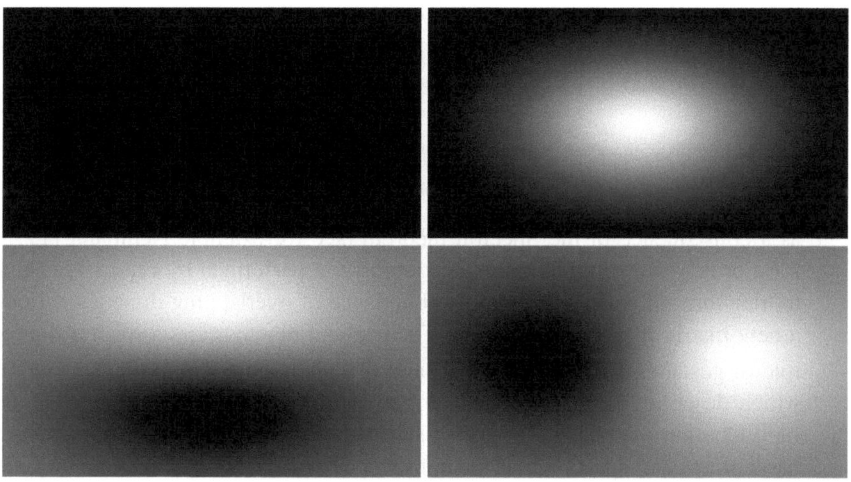

**Fig. 3.10** The paradoxical background (*top left*) and the three main eigenvectors of its connectivity field are shown (*top right* and *bottom*). The eigenvectors are obtained by numerical simulation. They corresponds to the 3 main formemes of visual semiotics: centrality/perifery, verticality (*top/down*) and laterality (*left/right*)

$$A_{ij} = \exp -\mathcal{D}_e^2(p_i, p_j)$$

$$(p_i, p_j) \; in \; \Omega$$

where $\mathcal{D}_e$ is the Euclidean distance.

The values $A_{ij}$ are now the entries of the affinity matrix playing the role of the preinidividual field. See Fig. 3.9.

By spectral decomposition of the affinity matrix $A_{ij}$ we obtain as principal eigenvectors the functions visualized in Fig. 3.10, well representing the structure of oppositions of the paradoxical background as analyzed by the Group $\mu$.

### 3.3.3 The Reduced Space as Spectral Embedding

The principal eigenvectors $x_i$ have a number of possible roles and provoke different kinds of individuations.

(1) Individuation of forms: The most salient eigenvectors $x_i$ correspond to *gestalten* (in the german meaning of good forms) and more generally to visual formants. Co-occurring features are mapped to the same eigenvector; features that do not co-occur are mapped to different eigenvectors. The eigenvector with the highest singular value represents the most important vector in the data (i.e. the vector that explains the highest variance of the matrix); the eigenvectors linked to the second highest eigenvalue represent the second most important vector (orthogonal to the first one), and so on.

(2) Individuation of relational graphs: The first rescaled eigenvector $\phi_{1i} = \sqrt{\lambda_1}x_{1i}$ build the rank 1 matrix $\hat{A} = \lambda_1 x_1^i x_{1i}$, that is the best approximation of the original matrix $A$ in the sense of least squares

$$min_{\hat{A}}||A - \hat{A}||$$

as a consequence of the Spectral theorem (see for example Neumann (1998) for a proof). The singular value decomposition performed by Eq. (3.4) is able to transform the matrix $A$, with an abundance of overlapping dimensions, into a new matrix of small dimension. Then salient eigenvectors $x_i$ play a double role of gestalten (in the sense of (I) and of salient relational graphs $\hat{A} = \lambda_i x^i x_i$ which describes the most salient properties of the original graph.

(3) Individuation of spaces: The set of salient rescaled eigenvectors $\phi_i$ play as a basis of a vector space, that is the consistency plane. The $\phi_i$ constitute the axes of the reduced space. Any image $I$ is then projected $\langle I, \phi_i \rangle$ in the space of plastic formants as a point whose positional value correspond to his characteristics of centrality, laterality, elevation.

In the following section the reduced space will be equipped with suitable potentials and it will be considered as the *space of control* of the classical morphodynamic structuralism, allowing for the categorization and the emergence of sign.

## 3.4 The Introduction of a Semiotic Extension

Before introducing the saussurian theory of sign, its main concepts and to its functional architecture as a natural semiotic extension of forms obtained through an harmonic process, we must focus on the principles of such a semiotic extension. We have stated earlier that this extension can be called natural in the sense that it

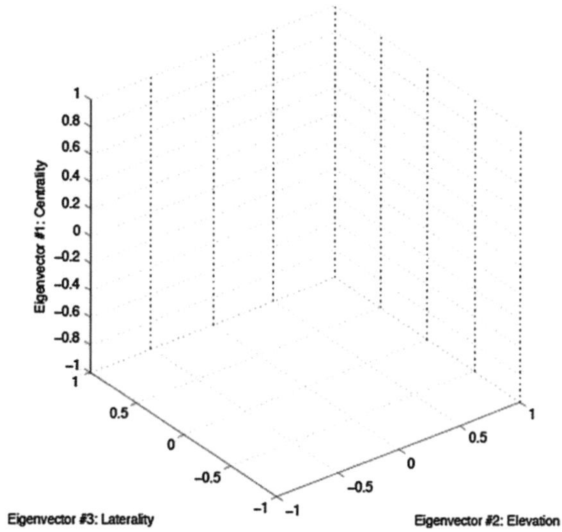


**Fig. 3.11** The set of salient rescaled eigenvectors $\sqrt{\lambda_i}x_i$ constitutes a reduced vector basis defining the consistency space as a spectral embedding

continuously connects the two levels of forms (harmonic and semiotic). Effectively, we can notice that this extension (i) prolongs and fulfills the proto-semiotic nature of the gestalten, and (ii) as it is based on a component resulting from the harmonic process, this component being precisely the saliency space built on eigenvectors, this extension ties the forms of the proto-semiotic and of the semiotic level. The first point (from proto-semiotic to sign) is rather clear and does not urgently require explanation. But the second must be examined in further detail, because the functional raising that is realized when conferring a control function to the axis carried by eigenvectors has an impact on the two levels result of the extension, this impact must be examined and clarified (Fig. 3.11).

The impact of such a functional extension is as follows: on one hand, the saliency space receives a semiotic content; and on the other, the degree of freedom of the semiotic structure decreases as it is anchored to the saliency space. The first impact is a current process of semiotization by adjunction of functional structure, and it does not set principles difficulties. However, the second impact introduces strong restriction on the instantiation of differential values (which are partly determined by the control axis). Seeing as these restrictions counter one of the most important principles of the saussurian theory (the arbitrarily principle de Saussure (1916)), they must be discussed.

To introduce our discussion, let us first recall that the saussurian arbitrarily principle does not essentially concern the connection between the two sides of the sign, but rather the place where a boundary can be located in a continuous substratum (amorphous mass of ideas or of sounds (de Saussure 1916, p. 156) producing differential values of content or expression. This principle asserts that the space where content or sound is located (substratum space) does not contains any data or information that could affect the instantiation of semiotic differential structures (the boundaries). Seeing as these emerging boundaries are carried by control axis, these axis can be freely situated in the substratum space. What is at stake here, is the largely debated question of the relationship between cognitive and semiotic categorization, and more generally between cognitive and semiotic processes. For some (cognitive linguists), semio-linguistic and cognitive processes share routines that make them compatibles. For others, cognitive and semiotic categorization are independent and therefore, in order to create a semio-linguistic description of the experiential world, one must conform them to each other. The limitation of these positions is that, in the former, the specificity of cognitive and semio-linguistic frames is negated, and in the later, the compatibility of resulting categorizations of cognitive ad semio-linguistic processes is not intrinsic, it is the result of a more or less mutual adaptation process.

One great contribution of the harmonic approach is designing a form of compatibility between semiotic and cognitive organization while keeping their specificity. The key to this solution can be found in the continuity established between the gestalten and semiotic structure via the space of saliency combined with the control function. From the harmonic perspective, this means that favored axes that sustain the process of semiotization exist. Therefore, semiotization appears as a natural extension of a given underlying cognitive process, such as perception. This explains the compatibility of perception and semio-linguistic categorization, while keeping their specificity.

## 3.5 The Functional Architecture of Sign

We have stated earlier that the semiotic extension of the harmonic process is obtained by adjunction of a control function on the saliency axis. Therefore, it makes sense to introduce the functional architecture of saussurian sign on the basis of its control component, and discuss the importance and the effectiveness of this notion in the semiotic field. The concept of control must be applied to semiotic field for at least three converging reasons. The two first reasons are mathematical and functional (related to the architecture of sign which, as we will see, includes a control component), and take place in the structuralist framework. Consequently, seeing as they depend on a particular semiotic theoretical approach to objectivity, these first two reasons can be seen as little weaker than the third one, which relies on a phenomenological analysis of the semio-linguistic sign. However this last reason, as it will logically lead us to introduce a structural perspective, is not fully disconnected from the first two. We will therefore begin by exposing and discussing the phenomenological reasoning behind the notion of control, the presentation of the first two reasons will naturally follow. The functional reasoning will be described through the exposure of the functional architecture of saussurian sign.

### 3.5.1 The Phenomenological Argument

At the very beginning of his phenomenological analysis of linguistic sign (developed from First Logical Research, up until Lectures on the signification theory), Husserl introduces the distinction between the expressive sign and the indicative sign. These two kinds of sign are respectively close to what the stoics call, with a partly-inverted terminological choice, the indicative sign and the commemorative sign. The stoics indicative sign rest on a duality whose second term cannot be thought of independently from the first one, in contrast with the commemorative sign whose second term can be known independently of the semiotic implication which connect the two terms composing the sign. Husserl uses the stoics distinction. He defines the indicative sign as a sign that functions with communicative logic, which is the transmission of information. The indicative sign coordinates two independent moments of awareness: there is first a certain state of awareness, which initiates with perception of the concrete symbolic data; and then, by its proper function, the concrete symbolic data draws awareness to another object which is the idea, the thing, or the state of things to communicate. Inversely, the signitive sign engages a single state of awareness: the perception of the sign as concrete data and the conception of its content are realized in the same awareness state. More precisely, the phenomenological identity of the expressive sign is not concrete perception which, following a communicational logic, would further lead to the actualization of a mental object. Inversely the mind turning toward content establishes the phenomenological identity of the expressive sign. Thus, the true sign, that Husserl calls expression, includes in its own phenomenological nature: in its own presentation, let us say, in its being there, the motion of consciousness toward a signification. What phenomenological analysis of semiolinguistic sign teaches us, is that the perception of sign is not a pure vision of some ideal object, it also includes concrete perception. When my eyes look at a written word, of course my mind is drawn towards the meaning implied by this word, but this meaning is not alone in the whole field of my consciousness, some concrete data is also present, which remains present, not at the same level as the meaning, but still there, as a kind of low level presence. To account for this phenomenological description of sign perception, Husserl (1995) introduces a hierarchy of attentional levels. Inside this hierarchy he retains the thematic and the primary levels. In this framework, the object of meaning belongs to the thematic level, in the sense that the awareness lives in and inhabits this object of meaning. The concrete perception of the sign stays at a simple primary level, which is the level of objects that are fully distinguished, but which do not belong to the central interest of the living mind. The objects of the primary attentional level are fully constituted in their perceptive design (as opposed to objects in the background whose shapes are not well-defined), but their presence and their actuality are not important to the living mind. Semiotic perception appears complex: it articulates a thematic object (the meaning) and a simple primary object (the sign as material data). However, what must be stressed here, is that the relationship between the thematic and the primary objects (between what Saussure calls the signified and the signifier) is not causal or logical. It is not like the connection between the two components of the indicative sign. It appears that the phenomenological approach of sign leads us to introduce a kind of relationship between objects which is different from a logical or causal connection. Different precisely in the way that the connected objects (i) must not be inscribed in any kind of consecution (temporal or inferential) and (ii) must have different ontological status. In other words, the type of relationship which is needed to account for the phenomenological description of the internal sign form is such that (i) it must distribute its own terms on different ontological strata and (ii) it must maintain the coexistence of the terms, though at different levels of awareness. The relationship of control in a morphodynamical framework satisfies precisely these two conditions: the control parameters coexist with the forms they control; and they don't share the same ontological status. Unlike the causal or the logical relationships, which respectively connects material or propositional objects, the control relationship establishes a link between objects that belong to different ontological strata, for instance, between an acoustic input and a perceived phoneme, or between a molecule and a morphology. Concerning the internal structure of sign, the notion of control connects symbolic data perceived as a material object to an intentional object (of meaning) in a single semiotic identity. The material data becomes significant through its functional role of control. This gives it a particular phenomenological profile, part of a whole, in a whole constituted by the controlling and the controlled objects, in such a way that this part is perceived as secondary to the composition of the whole, the thematic level being occupied by the controlled object. The notion of control appears thus to be a sufficient concept to account for the phenomenological structure of the sign that appears to connect its two sides in a dissymmetric whole. And thereby, it throws us directly into a structuralist issue, for the simple reason that, as Petitot (1985, 1992) has stressed, the control is the realization of the category of causality in the structuralist ontology. So, it is logical to extend our discussion on the concept of control to the structuralist framework. Precisely, we will have to show that this concept works with the architecture of sign developed by Saussure (de Saussure 1974, 2002), and further, that inside this structural framework, it provides the key to a mathematical characterization of the category of difference (Fig. 3.12).

### 3.5.2 The Mathematical Argument

Let us first recall that, in Saussures mind, semiotics unities and identities are purely relational, and that the kind of relation involved in this structural ontology is that of difference. Before discussing this kind of relationship, we must stress that unlike for logical objects, in this case relational conformation concerns unity as a well as identity. In fact, logical objects are built on the a priori category of autonomous unit (cf. Curry, who calls it unspecified objects (1963)) which can be indicated through the choice of particular symbols (letters X, Y… for instance). However, those single

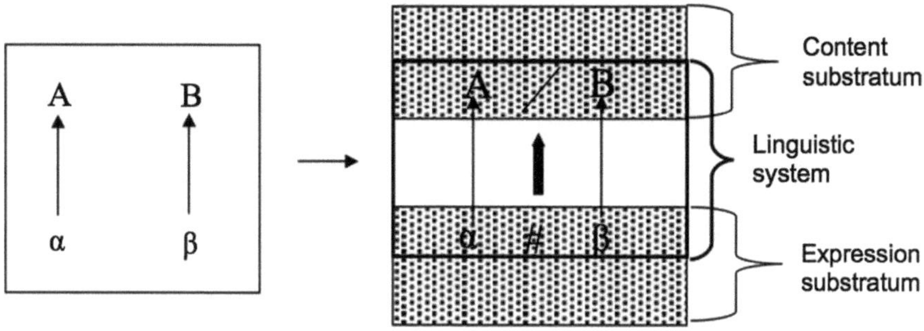


**Fig. 3.12** From indicative sign to expressive sign

units, which contain their own unity, do not possess a specific identity (they are mutually indistinguishable), but receive it through the formal relationships they contract with each other (for instance R(X, Y)) as established in the axiomatic part of a formal system. Unlike logical-symbolic units, semiotic units do not exist, regarding their identity as well as their unity, outside of the relational system they constitute and belong to. This relational ontology is fully realized by the relationship of difference in its topological meaning. The underlying thought process is that of a continuum in which a boundary draws two neighboring sides, and thereby produces a categorization of this continuum. Clearly, these two sides are absolutely correlated to this boundary: if the boundary disappears, the two sides will also disappear. So, differential units are fully relational units, they do not exist, in terms of individuation, outside the differential connection on which they are built. Before discussing the question of differential identities, and the architecture of saussurian sign, we would like to quickly sketch the mathematical reasoning behind the concept of control. As indicated by Petitot (1985), one of the core issues in the structural and morphodynamic approach is that of the categorization of a continuum (the substratum space), that is the categorization of a set of units which share the same qualitative identity, yet do not themselves contain any principle of differentiation. To by-pass this limitation, the initial set is placed inside a more complex set of terms (a dynamic space) which provides possible categorizations. And to make the feed-back of categorization (carried by the complex set) possible back on the substratum space, the connection between the later and the former set is thought of in terms of control. The units from the substratum space control the design of forms in the dynamical space, and inherit the qualitative variations of these dynamic forms through an instantiation of a boundaries network which produces a categorization of the substratum space (see upper part of Fig. 3.13). It appears that the theoretical choice of a differential relationship that creates categorical unities inside the continuum of a substratum, leads us to introduce the concept of control. In other terms, the control connection is (one of) the mathematical key that accounts for the categorization of a continuum through the emergence of a network of boundaries. This is exactly what the saussurian approach to sign needs, where similarly the problem is to account for the subdivision of a semantic or phonetic continuum: "Nous

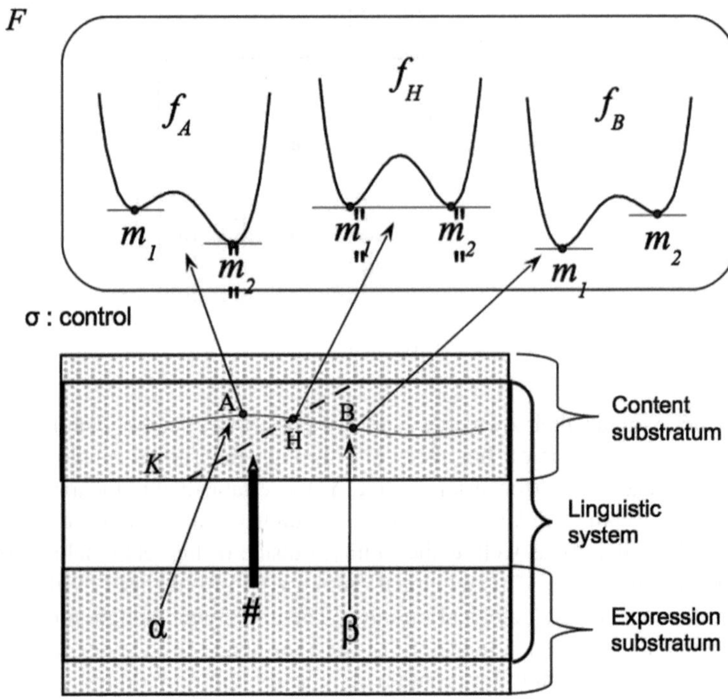


**Fig. 3.13** The morphodynamical solution to substratum categorization

pouvons représenter le fait linguistique dans son ensemble, c'est-à-dire dire la langue, comme une série de subdivisions contigues dessinées [...] sur le plan indéfini des idées confuses, [etc.]" (de Saussure 1916, p. 156). This is the second reason we use this functional notion. Let us now more precisely describe how the control connection operates in the Saussures theory of sign, and how it permits the design of a mathematical architecture of semio-linguistic sign. In what follows, we will quickly sketch the main elements of Saussures conception of sign that serve our purposes. A more complete presentation can be found in Piotrowski et al. (1997), Piotrowski (2009).

### *3.5.3 The Architectural Argument*

We have seen that a differential relationship creates fully relational units, and that Saussure uses this kind of relationship to define semantic identities. But in Saussures theory, the boundaries that delimit topological units in the continuum of meaning are not primary and static facts. The boundaries are only a resulting fact, which express the deeper relationship underling the process of semio-linguistic individuation. This relationship is an oppositive relationship where the various terms are in conflict for their maximum extension into the substratum space. The boundaries on which semantic terms are produced are to be seen as the resulting stabilization of the underlying and primarily conflict between them, a kind of competition for a maximal expansion of each. Saussurian structuralism is therefore topological and dynamic, its identities are fully defined through the principles of negativity and mutual limitation. The following quotation clearly sums up these concepts: "[…] tous les mots qui expriment des idées voisines se limitent récip-roquement" (ibid.: 160); [le sens de *loup* est d'être ce que *chien* nest pas (négativité) et] *chien* designera le loup tant que le mot *loup* n'existera pas [;] si par impossible on n'avait choisi au début que deux signes, toutes les significations se seraient réparties sur ces deux signes (Godel et al. 1969, p. 199). But Saussures differential units are not built on a pure formal process, as it is clear in the previous citation. In these statements (and in many others), we notice that with the term signification Saussure takes account of an access to the actual meaning of each sign. When talking about the meaning of *wolf*, he grounds on the various tokens of what is not a *wolf*. In other terms, the negative identity of *wolf is* not only the result of a purely formal process, but is also built on the possibility to access and precisely to look over partly the set of token which are the possible actualized occurrences of the words whose meaning is opposed to *wolf*, that is *dog*, etc. The negative meaning appears to be the articulation of a differential form in the substratum of content. The crucial question is then: how does Saussure conceptualize this kind of access to tokens (which is, lets say, half of the differential identity) of the substratum space? The answer is particularly interesting because it brings us back to the husserlian and stoic notion of indicative sign. In fact, in the beginning of the *Cours* (de Saussure 1916), Saussure criticizes the indicative conception of the sign, and calls the indicative relation between a symbolic unit and a conceptual unit the signification relation. In other terms, the signification relation (also called exchange relationship) is a connection that is established not between the two sides of a sign, but between the concrete symbolic data and some token of content substratum, for instance a concept, a thing or a state of thing. The issue then appears to be the problem of the articulation of the differential form with the signification relation. This is exactly the problem which lies at the core of the notion of value, and that Saussure did not believe could be resolved. Let us briefly recall here the definition of value: "[les valeurs] semblent régies par ce principe paradoxal. Elles sont toujours constituées: (1) par une chose *dissemblable* susceptible d'être *échangée* contre celle dont la valeur est à déterminer; (2) par des choses *similaires* qu'on peut *comparer* avec celle dont la valeur est en cause" (de Saussure 1916, p. 159) (Fig. 3.14).

Another way to tackle the problem of the articulation of differential form with the signification relation is to put it in terms of a promotion of the indicative sign up to a expressive sign (cf. supra). We have seen that the relationship of signification underlies indicative sign, and thus, the integration (as in the concept of value) of this relationship to differential forms must lead to the signitive sign. Before intro-ducing the solution to this problem, we would like to point out that, in Saussures theory, the differential relation between units of content (the signified) is not the same of that between the units of expression (the signifier). Whereas the signified

Fig. 3.14 Cusp singularity: the 2 dimensional space of stabilization

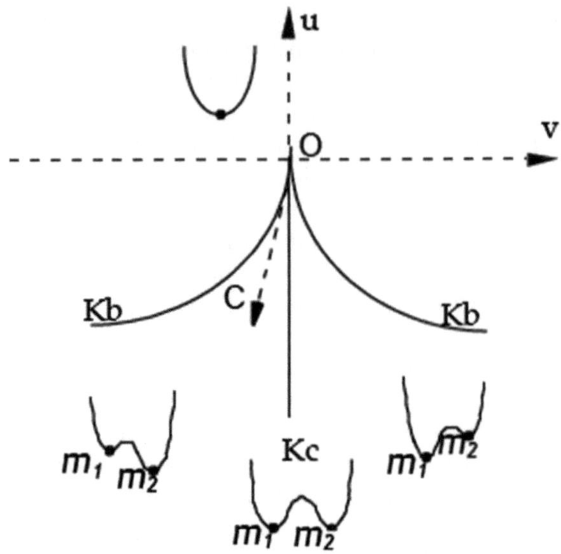


are built on and share topological and dynamical connections (the phonemes also), the difference between the signifiers is not an essential part of their identity, which is provided instead by their phonemes composition. "Deux suites de sons ne peuvent différer que par le nombre, la qualité et l'ordre des unités irréductibles" (Saussure in Godel et al. (1969, p. 199). From a formal point of view, the difference between signifiers is a relationship with a binary value: the same or not the same (compared with the semantic of the negative relationship: A is all that B is not). This kind of difference is called distinctive.

The problem we now have to treat is precisely the following: how to raise an exchange relationship (notated →) between symbols (notated $\alpha, \beta, \ldots$) and tokens of the substance of content (notated A, B, …) to an oriented and constitutive link between distinctive differences (notated #) and negative differences (notated /), for instance between *dog* and *wolf* as oppositional area established on each side of a boundary. That is: from $[\alpha \rightarrow A$ and $\beta \rightarrow B]$ to $[\alpha \# \beta \rightarrow A/B]$ (Fig. 3.14).

The solution is provided by the morphodynamical response to the categorization problem, that is: the boundaries that categorize an external space reflect the qualitative categorization of the dynamics (belonging to an internal space) that the external units control. We will have to introduce a dynamical space above the content plane and give its units the role of controlling the dynamics. In its simplest form, we obtain the following functional architecture:

On the upper part of this morphodynamical system, the tokens of content (A, B, …) determine dynamics $f_A, f_B \ldots$ (in a set $F$ of dynamics), here defined as potential functions, which establish the states of an internal system. The absolute minimum indicates the actual state of the system, while the state corresponding to the relative minimum is virtualized. In this picture the dynamic $f_A$ enters in opposition to the dynamic $f_B$ because its actualized state (given by $m_2$) is virtualized in $f_B$ while its virtualized state (given by $m_1$) is actualized in $f_B$. This expresses the fact that $m_1$ and $m_2$ are in competition for actualization. In the external space (the substratum of content) this competition for actualization is expressed by a (set of) boundaries. When following the path from A to B, in the first part (until H is reached) the dynamics corresponding to token of this path share the same qualitative identity, that is the same absolute and relative minima. On the other side of H, the dynamic scheme is inverted. Point H on the boundary is a catastrophic point, the dynamic it determines is unstable and its stabilization leads on one side to the actualization of $m_1$ (potential of the same qualitative type than $f_B$) and on the other of $m_2$ (potential of the same qualitative type than $f_A$). On the lower part of the morphodynamical device, the *exchange* relationship $\rightarrow$ (between $\alpha$ (resp. $\beta$) and A (resp. B)) is raised to a control function, which will be called primary. More precisely: the exchange relationship between an expression unit ($\alpha$ for instance) and its corresponding unit (here A) in the content substratum is extended from the content substratum W to F by using the field $\sigma$. In this manner, through the effect of the composition $\sigma o \rightarrow$, and relatively to the categorization process which is its structural outcome, the relation $\rightarrow$ receives the functional role of a control through a feedback logic.

Finally, in this kind of morphodynamic architecture of sign, the saussurian signifier is no more the substantial unit of the expression plane. However, it is this unit, as transmuted by the role of control it receives, and that orients it toward the differential (oppositive and negative) content identity it determines through an opposition (related by some boundary) to another signified. The holistic but dissymmetrical internal form of sign, as defined by Saussure, is well captured here. Let us now explore the connection between the saussurian sign and the salient space.

### 3.5.4 The Connection Between Gestalten and Sign

To connect the gestalten and the sign we have to introduce one of the main features of the morphodynamical approach, which has been ignored here in order to simplify our presentation. The dynamics which express conflict to actualization result from the stabilization of a more unstable one. In addition, topological distribution and the complexity of the stabilized dynamics is, under some conditions of structural stability, strongly constrained. This stabilized dynamic space is determined by the number of directions of stabilization (the control dimensions) through which the initial unstable dynamic is stabilized. For instance, in the simplest case, when the dimension of control (number of axes of stabilization) is 2 (axis $u$ and $v$), the stabilization space and its topology are the following (cusp):

For instance the axis $v$ represents the continuum of verticality and $k_c$ is the differential relation which establishes on each side the signified "high" and "low". As can be seen, the dynamics and their topological distribution (boundaries $k_b$ and $k_c$ between their different qualitative types) are determined by the parameters (u, v) of the control space. The differential organization of a substratum space will follow the same organization principle. It will be carried and determined on the basis of a system of control axes which define the directions of stabilization of the most unstable dynamic which is a kind of initial germ. It follows that the location of a system of boundaries in a substratum space (of meaning or sounds) that produces oppositional identities through a process of categorization, is entirely correlated to the location of the axes that control dynamical stabilization. Consequently also, the existence of some favored axis, such as the salient axis that emerge through a harmonic process, is the key to understanding the semiotic qualification of the data it refers to.

In fact, when the data set is approached at the microscopic level of its units and their mutual connections, it does not possess a favored orientation. It is an homogeneous and isotropic space, and its possible differential categorization is therefore free and arbitrary. But this is not the case when a higher level of its salient structure is considered. In that case, the saliency structure defines certain main axes which are naturally candidates for the categorization process that produces differential identities (the signified), and therefore, in morphodynamical terms, these axes are candidates to operate as control axes of an upper dynamical space. To conclude, the system of salient axes in a data set is a natural fulcrum of its semiotic raising and individuation.

## References


Arnheim, R.: Toward a Psychology of Art. University of California press, Berkeley (1966)

Bouquet, S.: La sémiologie linguistique de Saussure, in Langages, n. 107: 84–92 (1992)

Bressloff, P.C., Cowan, J.D.: The functional geometry of local and long-range connections in a model of V1. J. Physiol. Paris **97**(2–3), 221–236 (2003)

Citti, G., Sarti, A.: A cortical based model of perceptual completion in the Roto-Translation space. J. Math. Imaging Vision **24**(3), 307–326 (2006)

Curry, H.-B.: Foundations of Mathematical Logic. Mc Graw Hill, New-York (1963)

de Saussure, F.: Cours de linguistique générale, Édition critique par R. Harrassowtitz, Engler, Wiesbaden (1974)

de Saussure, F.: Écrits de linguistique générale. Gallimard, Paris (2002)

de Saussure, F.: Cours de linguistique générale. Payot, Paris 1972 (1916)

Deleuze, G., Guattari, F.: Milles plateaux, Les éditions de Minuit, Paris (1980). (Trans.: A Thousand Plateaus, University of Minnesota Press, 1987)

Deleuze, G.: Conference sur le temps musical, IRCAM. http://www.le-terrier.net/deleuze/19ircam-78.htm (1978)

Edeline, F., Klinkenberg, J.-M., Minguet, P.: Traité du signe visuel: pour une rethorique de l'image, Ed. Groupe μ, Seuil, Paris (1992)

Favali, M.: Master thesis, University of Bologna. Private communication (2013)

Field, D.J., Hayes, A., Hess, R.F.: Contour integration by the human visual system: evidence for a local association field. Vision Res. **33**, 173–193 (1993)

Godel, R.: Les sources manuscrites du Cours de Linguistique Générale de F. de Saussure, Genève: Droz 1969 (1957)

Hoffman, W.C.: The visual cortex is a contact bundle. Appl. Math. Comput. **32**, 137–167 (1989)

Hubel, D.H., Wiesel, T.N.: Ferrier lecture: functional architecture of macaque monkeyvisual cortex. Proc. R. Soc. Lond. B **198**(1130), 1–59 (1977)

Husserl, E.: Recherches logiques (prolégomènes), vol. 1. PUF, Paris (1969)
Husserl, E.: LeÃ§ons sur la théorie de la signification. Vrin, Paris (1995)
Husserl, E.: Recherches logiques (recherche 6), vol. 3. PUF, Paris (2000)
Husserl, E.: Recherches logiques (recherches 1 et 2), vol. 2, part. 1. PUF, Paris (1991)
Husserl, E.: Recherches logiques (recherches 3, 4, 5), vol. 2, part. 2. PUF, Paris (1993)
Husserl, E.: Thing and Space (1907)
Piotrowski, D.: Morphodynamique du signe, I Larchitecture fonctionnelle. Cahiers Ferdinand de Saussure **63**, 185–203 (2010)
Piotrowski, D.: Morphodynamique du signe, II Retour sur quelques concepts saussuriens. Cahiers Ferdinand de Saussure **64**, 101–118 (2011)
Piotrowski, D.: Morphodynamique du signe, III Signification phénoménologique. Cahiers Ferdinand de Saussure **65**, 103–123 (2012)
Khler, W.: Gestalt Psychology. Liveright, Bell, New York, London 1929
Merleau-Ponty, M.: Phenomenology of Perception. Routledge, New York (2012)
Perona P., Freeman, W.T.: A factorization approach to grouping. In: Burkardt H., Neumann, B. (eds.) Proceedings of the ECCV, pp. 655–670 (1998)
Petitot, J.: Les catastrophes de la parole. Maloine, Paris (1985a)
Petitot, J.: Morphogenèse du sens: 1, Pour un schématisme de la structure. PUF, Paris (1985b)
Petitot, J.: Physique du sens: de la théorie des singularités aux structures sémio-narratives. Paris: éditions du CNRS (1992)
Petitot, J., Tondut, Y.: Vers une Neurogeometrie. Fibrations corticales, structures de contact et contours subjectifs modaux. Mathematiques, Informatique et Sciences Humaines, 145, EHESS, CAMS, Paris, 5–101 (1999)
Piotrowski, D.: Phénoménalité et Objectivité Linguistiques. Champion, Paris (2009)
Piotrowski, D.: Dynamiques et Structures en Langue, Paris: Editions du CNRS (1997)
Sanguinetti, G., Citti, G., Sarti, A.: A model of natural image edge co-occurrence in the rototranslation group. J. Vision **10**(14), 1–16, 37 (2010)
Sarti, A., Citti, G., Petitot, J.: Functional geometry of the horizontal connectivity in the primary visual cortex. J. Physiol. Paris **103**(1–2), 37–45 (2009)
Sarti, A., Citti, G.: On the origin and nature of neurogeometry. La Nuova Critica (2011)
Sarti, A., Citti, G., Petitot, J.: The symplectic structure of the primary visual cortex. Biol. Cybern. **98**, 33–48 (2008)
Sarti, A., Citti, G.: The constitution of visual perceptual units in the functional architecture of V1 (submitted)
Scot, D.: Le principe d'individuation (De principio individuationis). Vrin, Paris (2005)
Shi, J., Malik, J.: Normalized cuts and image segmentation. In: Proceedings of the IEEE Conference on Computer Vision and Pattern Recognition, pp. 731–737 (1997)
Simondon, G.: L'individuation à la lumière des notions de forme et d'information, préface de J. Garelli, Paris, Millon (2005)
Wang, D.: Emergent synchrony in locally coupled neural oscillators. IEEE Trans. Neural Networks **6**, 941–948 (1995)
Weiss, Y.: Segmentation using eigenvectors: a unifying view. In: Proceedings of the IEEE International Conference on Computational Vision, pp. 975–982 (1999)
Wilson, H.R., Cowan, J.D.: Excitatory and inhibitory interactions in localized populations of model neurons. Biophys. J. **12**, 1–24 (1972)

# Chapter 4
# Morphogenesis Under Construction: Tracing the Process of Individuation Along Physico-Aesthetic Coordinates

**Claudia Mongini**

**Abstract** In this article I examine the concept of bio-info sensibility recently developed By Alessandro Sarti and Franco Berardi (Bifo) and frame it in the realm of Guattari's analysis about concerning techno-mutative conditions and Simondon's idea of technoaesthetics. These two concepts allowed to place idea of bio-info sensibility within the constitution of a morphogenetic field delimited by both microphysical and aesthetic conditions. I will subsequently depart by constructing the conditions for this morphogenetic field along lines of thought opened by Simondon, Deleuze, Deleuze and Guattari. The concept of asymmetry in its elaboration by Alberto Toscano serves as a juncture between the development of theoretical conditions of individuation and their pragmatics of realization. The elaboration of this point of view will subsequently allow me to move back to Simondon's theory in terms of a concrete research for possible components leading towards the morphogenetic construction of the concept of bio-info sensibility.

In a discussion about Simondon and his influence on Deleuze, Alberto Toscano has proposed to reevaluate the creative potential of both Deleuze's and Simondon's thought by delineating "a problem that exceeds them both" (Toscano 2009, p. 381). He individuates the field of interest in the question of "(post-)cybernetic heritage", a shared concern between both thinkers. Simondon's reading of cybernetics has lead him to the development of the notion of individuation, elaborated by linking reflections about technical objects to the question of ontogenesis (2009, p. 384). Through the notion of 'spatiotemporal dynamisms', Deleuze came subsequently to extend the topology of Simondon. A subsequent move of thought in its evolution with and against cybernetics developed out of the joint work of both Deleuze *and* Guattari, who developed the pragmatics of an "*immanent* construction of alternatives to axiomatic capitalism" (2009, p. 383) in the shape of an antistructuralist pragmatics.

Here, I propose one further step towards extending the problem which Toscano has identified in the frame of the post cybernetic condition. I intend to analyze this

C. Mongini (✉)
Université Paris 8, Saint-Denis, France
e-mail: cmongini@gmx.at

 
A. Sarti et al. (eds.), *Morphogenesis and Individuation*,
Lecture Notes in Morphogenesis, DOI 10.1007/978-3-319-05101-7_4

issue by departing from the concept of "bio- info sensibility" recently coined by Franco Berardi (Bifo) and Alessandro Sarti (Berardi and Sarti 2008).

> Sensibility can be understood as the *condition* of perceptibility of objects, phenomena, processes which have not yet emerged at the level of consciousness, and consequently do not possess a stable and recognisable form within the cognitive configuration. (69, my translation)" (2008, p. 69)

Sensibility develops within organic matter, but gets also specified and thus determined through the more abstract relation of information. And information is not a stable factor, but gets altered by cultural, technical and political phenomena.

The roots of this concept lie in the thought of Félix Guattari who in the 80ies has spoken of "the age of planetary computerization", emphasizing that technical development cannot be disentangled from deep forms of social and subjective mutation (Guattari 1989, p. 11). Guattari specifies this issue in a conversation hold in 1985 with Michel Butel and published in "Les Annes d'hiver" (Guattari 2009), in which he considers the problem of the socio-technical mutative conditions in relation to the question of producing a territory. Territory is understood as "the ensemble of the projects or the representations on which a whole series of behaviors and commitments will pragmatically unfold both within time and within social, cultural aesthetic and cognitive spaces" (133–134, translation mine). It gets thus extended from its pure geographical definition, towards the inclusion of mechanisms investing features of bodily production which are operative *before* the modality of subjectivity gets invested by consciousness; by furthermore investing a socio-cultural as well as an aesthetic dimension, it becomes the ground for an epistemic problem. In a subsequent conversation with Robert Maggiori published in the same book, Guattari introduces the concept of *intradisciplinarity* (168), a minimal machinic condition linking heterogeneous fields of knowledge production. This concept expresses the concreteness of production mechanisms, and is intended as a sort of *pragmatic* differential operator which enables the articulation of singularities relative to a specific domain and links them to other ones pertaining to different areas of knowledge.

The concept of bio-info sensibility grasps the minimal operative constraints entailed by the notion of intradisciplinarity, in terms of an *aesthetic* condition. The development of informatics and biotechnologies Bifo and Sarti argue, has protracted the mutations diagnosed by Guattari towards a new biomechanical (and political) state. Bifo ad Sarti's interest into the topic is not concerned with a merely sociological framing of the current development, but aims to perceive the intensive conditions demarcating the very roots of the problem (2008, p. 83). Taken from this perspective, Bio-info sensibility does not describe an already constituted development, but expresses the morphogenetic operation allowing for the intertwining of a series of complex and multiple fluxes connecting imaginary, technological, psychic and perceptive microstates (2008, p. 117).This flow of interconnection is what links the scientific and philosophical problem of emergence of form to a question of aesthetics.

Gilbert Simondon had previously linked the fields of technique and aesthetics by means of the notion of technoaesthetics which was used once in a letter addressed to Jacques Derrida (Chabot 2013, p. 141). There, aesthetics expresses the way a tool adapts to its function; it is grasped in its operational condition towards action, and denotes the upcoming of a surprising event, instead of the predictable and repetitive state, the condition characterizing the technocratic attitude towards technology (2013, pp. 141–142). Aesthetics becomes a bridge towards a specific coadaptation between conflicting experiences, acquiring thus both a functional and an operative character.

By means of the notion of Bio-info sensibility, Bifo and Sarti emphasize the aesthetic condition in terms of the pragmatics of its procedural and technological aspects and come to adopt an operational point of view which is similar to the one proposed by Simondon.

Their specific focus however, concentrates upon the material aspects of formation from the perspective of emergent microscopical features. Matter gets expressed from the point of view of its intensive condition, located in between the state of biological substrate and its physical operation of constitution. In concomitance with Deleuze's and Guattari's notion of Body without Organs, matter is not understood in terms of an object occupying a preconstituted space, but instead emerges *within* a space "to a given degree—to the degree corresponding to the intensities produced" (Deleuze and Guattari 1987, p. 278). The question arising from this problematic configuration, concerns the *technical* conditions which are necessary for a morphogenetic microphysics to become constitutive of an aesthetic realm, along the relation previously drawn by Simondon.

My approach to this question evolves in the following way:

I will depart by constructing the process of individuation from the point of view of its material (i.e. microphysical) conditions, by examining the Simondonian concept of disparation and its connection to the idea of transduction. I will subsequently extend the analysis towards the inclusion of Deleuze`s concept of dramatization. These morphogenetic conditions will be analyzed along the concept of asymmetry elaborated by Alberto Toscano; this concept shifts the theoretical conditions of morphogenetic individuation towards the possibility of pragmatic realization upon the plane of composition expressed by the joint work of Deleuze *and* Guattari. Following this point of view, I move back to the thought of Simondon and consider the concepts of information and image as (possible) constitutive components of the idea of bio info sensibility.

## 4.1 Conditions of Morphogenesis 1: Individuation, Transduction and Disparation in Simondon

"Qu'est ce que l'individuation?" Gilbert Simondon asks in the first chapter of "l'individu et sa genese physico biologique" (Simondon 1964, p. 26) entitled Forme et Matiere. This question is posed at the level of the consideration of the ambivalent limit between matter and form; it is fundamentally a question of morphogenesis, as it regards the genetic conditions of both formal and material development. The individual is considered as being the outcome of a process of individuation located at the boundary between the internal state of intrinsic formation of matter and extrinsic constitution of its surrounding milieu (Simondon 1964, pp. 67–69). This genetic and liminal character is constitutive of the status of interactive communication that the individual develops in its very process of becoming. The individual as such constituted is not a "being in relation" but "being *of* relation", the "active agent of a relation" (1964, p. 69). It develops in the tension between two separate degrees of reality, "two orders in a state of" disparation.

Simondon saw individuation emerge from a material plane of relations made up of differing potentials of energy that he called the 'pre-individual' state. Borrowing from quantum mechanics, Simondon argues that pre-individual disparity is at once quanta and wave, energetic and material, and these atomic and sub-atomic levels are directly expressed on a macroscopic and hence sensible level (Simondon 1964, p. 135, 2007, p. 108). Individuation in this sense is a partial and local resolution of these disparate energetic/material fields, one that does not exhaust its ontogenetic possibilities and therefore retains the potential for further development. The pre-individual disparity that is the condition of an individuation, and indeed that which individuation expresses, is what Simondon calls 'information', and as such it has no content, structure or meaning. 'Information' is rather an event of onto-genetic emergence (individuation) that cannot be thought separately from the pre-individual field that it simultaneously expresses and constructs (crf. Deleuze 1995, p. 246). Thus Simondon's idea of information, although grounded in the conceptional realm of cybernetics, has a more complex construction than the probabilistic formalism of information theory and cybernetics (Shannon and Weaver, Wiener), where information is inversely proportional to noise (crf. Simondon 1992, p. 304). Simondon argues that the process of individuation has two immanent dimensions, that of the individual in becoming, and that of its milieu, these two elements being at once divergent and inseparable within the process of individuation. This has an extremely important result for Simondon; 'It would allow us to grasp, on the macroscopic level, a phenomenon that is rooted in those states of the system belonging to the microphysical domain, molecular and not molar' (Simondon 1992, p. 304). In other words, an individuation and its milieu are reciprocally determining, each being expressions of information that construct each other.

In the introduction of l' *individu et sa genese physico biologique*, Simondon derives the concept of *preindividuality* out of the dualistic description of the atom defining it as a state "beyond unity and identity, something capable of being manifested as either wave or corpuscle, matter or energy." (1992, p. 302) The assessment of these different configurations allows for a continuous energetic exchange, leading the system into a state of *metastable equilibrium:* contrary to the stationary state of stable equilibrium, metastable equilibrium continuously allows for new processes of transformation to happen. This status entails the presence of a level of *potential energy* which constitutes the basis for the maintenance of metastability.

The notion of potential energy is the basic component of what Simondon calls the preindividual state, a material plane of relations out of which the individual emerges through a process of individuation. The relation of the preindividual to the individual is a complex and persistent one, which means that the genetic process not only constitutes the individual singularity, but accounts for its further individuation through the nature of its relations. Individuation in this sense, is a partial and local resolution of these disparate energetic fields, one that does not exhaust its ontogenetic possibilities, and therefore retains the potential for further development. Relations do not appear as a secondary process, consequent to the formation of the individual, but are part of the generative process itself. This gives rise to processes of 'resonance' between different levels of magnitude—both on the microscopic and macroscopic, individual and collective levels. In this way communication between different orders of magnitude can take place, a process Simondon calls 'transduction' (see Simondon 1992, p. 315). As Adrian Mackenzie has put it, 'Transduction is a process whereby a disparity or a difference is topologically and temporally restructured across some interface. It mediates different organizations of energy' (Mackenzie 2002, p. 25).

Sarah Margairaz (Margairaz 2010) observes that Simondon describes transduction in terms of the concept of intuition. Intuition is understood as a modality of thought which "can be applied to any domain transversed by a genetic operation, because it follows the *genesis* of beings, taking in this way every being at its level of unity." (Simondon 1992, p. 236, my emphasis). This means that the production of analogies between heterogeneous domains, can be thought only in terms of conceiving genetic relations between different processes of individuation and not in terms of establishing connections amongst pre-given structures (see Margairaz 2010, p. 9). Because of this strong relation to an ontological dimension of becoming, Simondon's idea of intuition maintains de facto a connection with substance; this material constraint accounts both for the creation of constitutive genetic relations, and for the genetic establishment of a precise limit to a "general spreading" of linkages. Transduction comes to describe a procedure that "while it may be applied to ontogenesis, it is also ontogenesis itself" (Margairaz quotes Simondon in 2010, p. 7).

By departing from a tension between heterogeneities which gets both constituted and subsequently resolved in terms of the production of a constituted individual, this procedure enacts a process of becoming. Anne Sauvagnargues notices that transduction is not only a physical operation determining a mutual transformation between structural and energetic conditions, but also a procedure of formation allowing for the theorization of changes in thought. The process of both topological and temporal dephasing intrinsic to the process, implies a new understanding of temporality in which creation gets intrinsically linked to a process of differentiation. By departing from the seminal level of a difference in intensity, a wider spatio-temporal field can be reshaped along divergent conditions (Sauvagnargues 2009, p. 62).

The process of transduction, explicating the dephasing of a novel configuration in space and time as well as a process of different material and energetic constitution, is thus strictly linked to the microscopic seed of disparation, which constitutes its catalytic event (Sauvagnargues 2009, p. 62). This idea has straightforward mutative effect upon the concept of the individual who

> neither unified nor identical, it becomes relative, phased, perpetually putting into play a process of individuation and an associated milieu. The individual is thus never relative to a *single* order of reality but is always transductive, implying a disparation between different dimensions, arising as the resolution of a problematic, a tension between disparates. [...] Furthermore, the individual never requires unity or identity, since it reclaims the heterogeneity of the phases from which emerges through disparation. (Sauvagnargues 2009, p. 63)

Having sketched the relation between disparation and its material conditions in Simondon, I protract the logic of this analysis along the lines thought produced by the Deleuzian concept of spatio temporal dinamisms.

## 4.2 Conditions of Morphogenesis 2: Dramatization in Deleuze, Haecceities in Deleuze and Guattari

In the fifth chapter of Difference and Repetition, Gilles Deleuze defines disparation as the unequal as such, "the condition of that which appears" (Deleuze 1995, p. 222). Disparation is conceived as the intensity of difference, a motor enacting a process of individuation. Along the pages of the chapter, the philosopher explains that the condition of disparation is enacted when disparate factors have found modalities of relation *in-depth* that is, modalities of intensive communication out of which qualified modalities of extension unfold to emergence. This is the onset of a process which is both individuating and dramatizing: the intensity of the originary disparity emerges into the formation of physical and sensual extensive qualities. Heterogeneous elements unfold into their actual space time conditions, and acquire a performative character, as their very modalities of constitution account at the same time for the creation of new relations and of the materiality of new deep communicative channels. In other terms: the resonance between different entities does not only open for new levels of communication, but lies at the very onset of their conditions of creation.

The term dramatization, before getting richly unfolded in Difference and Repetition, constituted the title of a lecture that Deleuze gave in 1967 (Deleuze 2004), at the French Society of Philosophy.

In this text, Deleuze reelaborates Simondon's idea of disparation within the concept of obscure precursor. The obscure precursor is in his own terms a abstract 'difference operator' which relates difference to difference. With the notion of difference operator, Deleuze transposes the concept of disparation from a physical level onto a more abstract mathematical level. In this way Deleuze sets the conditions to break up Simondon's topology still characterized by a sort of general uniformity due to its straightforward analogy with physical fields of potential energy. In reference to differential calculus, Deleuze poses the problem at the level of infinity. It is at the infinitesimal condition, that the intensive condition becomes tangible; that the *intensity* of disparity becomes indistinguishable from its *extensity* i.e. from its more proper physical and sensuous qualities.

The uniform field of energy becomes the theatre of "spatiotemporal dynamisms" (Deleuze 2004, p. 96).

The field splits up into a coupling between different series of singularities, multiple phenomena of internal resonance, and an inevitable movement in form of an amplitude, which give rise to a series of states of intensity. This microlevel of dynamical communication gets generatively individuated in the Signal Sign Systems, accounting for the "flashing" in between disparate order of singularities. As Alberto Toscano depicts it "a complex interpenetration of formation and functioning" (Toscano 2006, p. 181) is at stake, which fully unfolds in the theatre of indi/drama different/citation proposed in Difference and Repetition. Individuation becomes thus the moment of intensity that dramatizes the differential potential of the virtual and accounts for the creation of divergent lines of actualization.

The border between the individual and the collective that Simondon already questioned at the level of the oversaturated preindividual, the generative condition of both individuality and collectivity, becomes further disgregated in the Deleuzian move: we assist here to a full break with the dialectical opposition between the one and the many and the consequent opening towards the dynamics of a theatre of multiplicity.

This ontologic (or more properly) ontogenetic move acquires a level of pragmatic consistency, in A Thousand Plateaus, book which Deleuze has written together with Félix Guattari (Deleuze and Guattari 1987). There, the complex topology of spatiotemporal dynamisms, gets materialized in the dynamics of the *haeccity* of libidinal fluxes and the preindividual field acquires the contours of real speeds and slownesses:

> The plane of consistency knows nothing of substance and form: haeccities, which are inscribed on this plane, are precisely modes of individuation proceeding neither by form nor by the subject. The plane consists abstractly, but really, in relations of speeds and slownesses between unformed elements, and in compositions of corresponding intensive affects. (Deleuze and Guattari 1987, p. 507)

A cross reading between Deleuze's solo texts (Difference and Repetition in particular) and the texts written with Guattari (with emphasis of A Thousand Plateaus), allows to pose the question of materialism in terms of the constitution of a relation between the abstract level of *genetic* dramatization and its *pragmatics* conditions of realization. Seen from this perspective, the intensity of disparation sets the conditions of an onto-epistemological problem, as it accounts for the *actualization* of the relation between both conditions of creation and modalities of metastable communication. It is a question of production of pragmatic formations of space and time, *and* of epistemic creation as well.

By tracing this line of thought, I hope to move one step forward towards the clarification of the relation between the physical process of morphogenesis and the pragmatics of bio-informational aesthetics. I follow Alberto Toscano's operation of placing Simondon`s process of individuation within the dynamics of an asymmetrical field of forces. An analysis of his *modus operandi* constitutes the content of the next section.

## 4.3 From the Spatio-Temporal Drama to the Pragmatics of Realization: The Concept of Asymmetry

The book "the theatre of production" (Toscano 2006) is centered upon the investigating the ontology of individuation from the point of view of onto-hetero-genesis, that is by departing from the examination of how the different both emerges from and accounts for the production of another different. This perspective Toscano argues, is constructed upon a fundamental asymmetry, which arises from the condition that the field of individuation cannot be determined by properties of already constituted individuals, as the process of individuation itself would be preempted from its process of becoming. Rather, it is the *operational* character of "unequal tendencies and disparate relations" (2006, p. 14) which accounts for the unfolding of a process of individuation.

This asymmetry acquires the symptomatic character of an anomaly in the stage when the question of the genesis of the individual is not anymore posed in terms of an abstract plane of relational possibility, but enters into stage of concrete *realization*. In this realm, the question of individuation becomes the question of an ontology of anomalous individuation, and acquires crucial importance when the theory of Deleuze is set in conjunction with the cartographies of Guattari.

The question of *heterogenesis* involves an affirmation of difference in the procedural constitution of univocity: Being is differential production, and the task of philosophy consists into a dramatization the tension leading towards a process of individuation. This assertion poses the problem of individuation in terms of a condition of organization; its *anomaly* lies in the deep cor-relation between an abstract operative process—the process of individuation *qua* process of heterogenetic formation—and the autonomous organization of matter. The question regarding the production of being gets thus reformulated in terms of the question of material self-organization. In a critical discussion of Kant's opus postumum, Toscano locates the problem of organization of being "in the passage between the realm of preindividual free matter (materia soluta) and that of bounded individuals or physical bodies (materia ligata)" (2006, p. 55).

It is in the realm of this constitutive tension between two different material states where the concept of disparation gets located. The event of individuation emerges out of a "non relation of disparation" (Toscano 2006, p. 145); disparation is resolved through the invention and the subsequent establishment of a local relation not existing before. In the fifth chapter of Difference and Repetition, Gilles Deleuze resolves the problem concerning the not (yet) relational character of disparation by unfolding it along the process of indi-drama-different/ciation: heterogeneous energetic differences of intensity unfold into the creation of actual conditions of space and time, and get dramatized in the process of *establishing new relations* as well as in the material constitution of new spatiotemporal dynamisms.

Dramatization as an abstract process, gets cracked by the immanence of the real by means of the encounter between Deleuze and Félix Guattari. This encounter produces the prerequisites for individuation to be "thought in terms of *haeccities*

[…][both] kinetically and topologically, as a concretion of speeds and affects; not hylemorphically (like the organism) in terms of the in-formation of matter." (Toscano 2006, p. 182). An asimmetry enters into the game, because this very movement of concretization accounts for a change in kind, a new dimension which cannot be permutated or brought back. It is an asymmetry which is tightly bound with the transformatory potential of practice: this is the anomaly of individuation (Toscano 2006, p. 184).

It is at this stage, that Toscano comes to re-design the virtual in terms of a stronger materialist condition than Deleuze himself does, by means of a cross-reading operation between Deleuze and Badiou. Following Badiou's critique, Toscano proposes to move away from the Bergsonian question of Time as the sole ground of the virtual whole (i.e. Bergson's paradoxes of time) and to consider instead the *differential production* of space-time, (i.e., Deleuze's uptake on Simondon), as the core motor not only of individuation, but also of the virtual itself. This means that both an energetics of disparity as well as the unfolding of spatiotemporal dynamisms, become the basic conditions by which the virtual gets understood *as* transcendental materiality. This is why, Toscano argues, Deleuze theory accounts for a radicalization of Simondon, as it demands from thought "an inhuman transformation a deindividuation of its own of the kind invoked by the 'dissolution' of the self and the 'fracture' of the ego" (Toscano 2006, p. 197). In Toscano's view, this move comes closer to the creation of an 'ethico-aesthetic experiment' rather than the development of a strictly ontological investigation.

Toscano's asymmetrical movement of thought enables the conception of the process of morpho-genesis not only in terms of its (micro)physical conditions, but also from the point of view of the *aesthetics* of an experiment (and of their inherent ethical implications).

It is by departing by the problematic field entailed in this delineation of asymmetrical effect, that I move back to Simondon's theory and come to address the specifics of the concept of *information* from the point of view of its constitutive conditions. By means of this examination I hope to sketch the conceptual frame for a minimal machinic condition allowing to tie the physics of a morphogenetic process with the pragmatics of an aesthetic operation. More specifically, I intend to explore the necessary constraints in order to address Guattari's concept of intra-disciplinarity introduced before from the point of view of its genetic conditions of emergence.

## 4.4 Information as Singularity

Simondon introduces the notion of information in his treatise, "Du Mode d'existence des Objects techniques" (Simondon 1958), book which is focused onto questions regarding the very nature of machines, their mutual relationship and the relationship with humans. By considering the principles of cybernetics in relation to more general physical laws, like thermodynamics, electricity principles and quantum mechanics, he elaborates a concept of information understood as a metastable state between the completely undetermined noise and the strictly fixed form. Information carries, according to its nature, determination to the system. On the other side, in the absence of any indetermination margin within the system there is no variation allowed, and thus no new state can be brought to the system: information thus ceases to have its effect.

That is, information is by definition unforeseeable by the system it relates to, but as a result, it leads to regularity and periodicity, that is, form (Simondon 1958, pp. 252–253). In accordance with gestalt theory, Simondon understands form as an upper limit to the acquisition of transmitted information (Moles 1971, p. 110). What thus demarcates the difference between this theory and the psychological theories of perception, is the importance the French thinker assigns to the aspect of dynamics. Form is not an entity necessarily defined a priori,—as it is conceived by Gestalt theory—but it is shaped, destabilized and reshaped through the information flow itself.

Due to this procedural character, information cannot be abstracted to an ideal value—an informational correlate of the Gestalts of the square or the circle—but is always dependent on the circumstances of the single context, the single experiment, that is, the field of forces given by the context. More than being determined by pure —abstract and context independent—quantitative or qualitative values, the reality of information is better captured by the measure of intensity. This concept is not descriptive for a predefined pattern of organization, but quantifies a tendency towards organisation in relation to the general degree of entropy of the system. It is not the point specific property that is taken into account here, but the general state of an area, as is the case in the study of electric or magnetic fields.

The systems in question here are characterized by states of disadaptation arising when information reaches a critical state. This means that at some crucial stage the general configuration of the system is affected, and it is reshaped towards a different, new state. Disadaptation is thus inherently different to the notion of degradation, that instead designates processes of decay.

In his construction of the concept of information Simondon doesn't only draw on ideas of Cybernetics, but also refers to Quantum Mechanics. It is Plank's 'quantum of action' and Heisenberg's 'indetermination principle' that allow him to construct a concept of information no longer bound to an abstract probabilistic formalism, as it was for cybernetics, but which instead is intimately related to the energetic conditions at the core of the material substance. As a result, indetermination is no longer conceived in opposition to information, but as intrinsically bound to it, as a quantity which is directly responsible for the process of transmission of the signal, rather than opposing it.

In the introduction of his subsequent book, "L´individu et sa genese physico-biologique" (1964) Simondon defined information as "the *signification* that emerges, when a process of individuation reveals the dimension through which two separate realities together become a system." (Simondon 1964, p. 311) Here, as well as in the preceeding treatise of 1958, the quantum mechanical perspective allows Simondon to rethink the cybernetic definition of information neither in terms of *pure abstract* mathematical quantities nor in an anthitetical relation with uncertainity, as it was the case for initial system theory, but in relation to differentiated *energy levels.* That is, information is reconnected to the basic atomic structure out of which it is generated. In this way, Simondon pursues a first step towards conceiving information in terms of its underlying material level.

By inserting the principle of quantum mechanics into the equation, Simondon makes a step towards understanding the meaning of indetermination from a broader perspective than the cybernetic point of view, which remains inherently bound to a positivist angle. The approach towards physics allows Simondon to conceive information as metastable, that is, as being *intrinsically bound* to indetermination and not in opposition to it, as it has been the case for cybernetics. While the French thinker agrees with cybernetics that information is set to carry determination to the system in question, by referring to quantum mechanics he examines information out of its own limit point: in the absence of any indetermination no new state can be brought to the system and information ceases to have its effect.

However, in the fourth section of the second chapter, named "Topologie, chronologie et ordre de grandeur de l ́individuation physique" the panorama changes: Simondon presents indeterminism as a quantity intrinsic to physical reality itself. "Indeterminism is not solely bound to measurement; its presence is also due to the fact that physical reality entails different degrees of magnitude which overlap each other, both topologically and chronologically, each of them being bound to its own specific becoming" (Simondon 1964, p. 125, my emphasis).

In the moment in which indeterminism acquires the status of an intrinsic feature of matter, the relation between the process of individuation and the preindividual becomes no longer an issue of return to the initial conditions, but is understood as a *coupling* between different entities in continuous transformative stages. The preindividual sets the conditions for the genetic becoming of the individual, and on the other side, this very process of individuation shapes the *chrono-topological conditions of the preindividual.* A complex mechanism of subtle resonances takes the place of the one-to-one relation between the individual and the surrounding field.

This new frameview introduces information as well as indeterminism as both strictly interconnected to the topo-chronology of the system. More specifically, indeterminism becomes the quantity able to delineate the mutual degree of separation between space and time components. It is in this section that a concrete step towards the reconnection between information and matter gets tangible.

As a consequence of this shift in perspective, the relational aspect of information becomes pivotal.

Relations do not appear as a secondary process, once the shape of individual is constituted, but are part of the generative process itself. "The relation to both the world outside and to the collective, is in fact a dimension of the individuation in which the individual participates due to its connection with the preindividual reality that undergoes gradual individuation" (Simondon 1992, p. 309). The construction of such a web of relations generates a flow of subtle communications between structures of different degrees, giving rise to processes of resonance amongst levels of different magnitude—both microscopic and macroscopic. By introducing the concept of relation and resonance, Simondon incorporates physical principles concerning the exploration of transitional laws between different orders of magnitude. This concern about transitions between micro and macrodomains demarcates the change in perspective about indeterminism. Far from being only bound to an external source of measurement, it plays the role of a pivotal key intrinsic to the system in question, by enabling relations between heterogeneous and structurally different macro- and microscopical levels.

It is at this stage that information acquires the status of a "singularity" in its dependence upon the *chronology and the topology of a system;* it will be able to create communicative levels between different degrees of chain reactions and becomes a "fundamental entity of individuation, which can be conceived in a topological or chronological dimension" (Simondon 1964, p. 127).

By departing from the material aspects of information understood *as* singularity, I intend to address the specifics of a pragmatics of intradisciplinarity which takes into account the asymmetric perspective of an aesthetic field. For this purpose I examine the concept of image developed in the recently published lectures on imagination and invention, which Simondon held between 1956 and 1966.

## 4.5 Towards an Aesthetics of 'Image'

In *imagination and invention* the concept of image in Simondon is conceived as an "intermediate reality between the abstract and the concrete, between the self and the world" (Simondon 2007, p. 13). Simondon explicitly affirms that its reality is not only that of speculative thought. Images materialize themselves, become matter. The image materializes both in the synaptic circuitry of the brain and in the flow of circulating information.

It is at this inframince level of materialization between internal and external circuits, that the activity of imagination sets on. This perspective extends the potential of imagination beyond acquisition of consciousness, as organisms are not only explicated in terms of their discernibility and perceptibility, but in terms of their development. The act of imagination, according to Simondon, comprises the process of reincorporation of phenomena within a process of becoming. This multifaceted intricate process of recomposition, is the constitutive condition of invention. In order to *invent the reality* of the object, a plurality of analytic viewpoints, philosophical, scientific, technical, aesthetic, (etc.) are necessary. The invented object becomes an "*actualized juncture*" (noeud d'actualité), resulting from the integration of different realities. Invention, more generally speaking, can be considered as a change in the organization level of the connected cycle of (external and internal) images. This change involves three levels by which the image is formed: a primary *biological* level which configures perceptual experience according to primary needs and fears. A secondary *psychological* level, implying a more specialized involvement of the nervous system, in form of motivation, conscious anticipation and desire. And finally, a third r*eflexive* level which shapes the image towards unity, by creating dynamical forms of resonance between internal states and the information deriving from the surrounding milieu.

This is the moment in which the *apriori* activity of image formation, involving the process of image constitution, is brought into a resonative condition with a constituted *a posteriori* stage of formed image. It is at this stage in which a wider cultural dimension gets interlaced with biopsychic processes (2007, p. 23). Problematic situations, that is, situations in which action is dualized, cut and segmentated, in which gaps and incompatibilities come to emerge. These can be obstacles but the means of invention, i.e. the re-installation of a level of compatibility between a milieu and its micromodalities of action (2007, p. 139):

> The invention establishes a certain type of return action, a feeedback which relates the complete result with the organisation of means ad its subensembles along a modality of compatibility. (2007, p. 140)

This problem affects the subject in its development: it is resolved when a form of communication gets established between the systems of action in which the subject is involved, and the reality of the result. Significantly, the subject is part of the degree of reality in which the result is achieved, and not of an imagined action. Reality is constructed in a twofold way: it involves the interaction between a milieu leading towards a final aim, and a field of accumulated experience. Invention thus involves a form of mediation, the research of a compatible solution within the current problematic situation and a synergy of past efforts. A variety of accumulated forces, needs to be reorganized and recomposed along the condition presented by the actual task (2007, p. 151): the proximity of a task establishes the presence of a gradient field, an energetic flow accounting for a mental recomposition of images deriving from past experience.

Departing from this physiological idea, invention can be extended to epistemic fields (such as the artistic field), this becoming the assessment of a mode of compatibility between formations and styles isolated in a previous stage.

The concept of image thus acquires the status of a 'quasi-organism', or 'living being'. Image develops within the process of a subjective formation, but due to its complex interrelational formation, it is relatively independent from the activity of unified consciousness (2007, p. 9). The production of image concurs towards an autonomous and self-organizing process of assemblage constitution, which encompasses a multiplicity of informational sources. 'According to this hypothesis,' Simondon argues, 'it would be possible *to consider every genuine relation as having the status of a being, and as undergoing development within a new individuation*' (Simondon 1992 p. 306). Image acts thus as a transfer between the re-processing of external information by the brain, and the creation of an aesthetic 'object'. Therefore, this notion provides us with the mechanism by which to extend the physical and neurophysiological functions of information towards the problem of aesthetic individuation.

## 4.6 Morphogenesis Versus Bio-info Sensibility

Along the development of this article, I have framed the concept of bio info-sensibility by the sketch of some of its (possible) components. I traced its roots along the concept of intradisciplinarity developed by Guattari, and Simondon's idea of technoaesthetics. These two latter concepts allowed to place idea of bio-info sensibility within the constitution of a field of forces delimited by the operative conditions of a micro physical transfer and the adaptive field of aesthetics. Its genetic conditions of emergence were determined by the examination of a series of concepts derived from the philosophy of Simondon, Deleuze, Deleuze and Guattari. Alberto Toscano's reading of Deleuze's concept of asymmetry, allows to translate the level of the physical genesis in terms of the pragmatics of a realization. It is by departing from this perspective, that I move back to Simondon's theory, from where I extract the concepts of information and image. Information is understood as the condition of singularity which changes the whole chrono-topological dynamics, while image acts as a transfer condition between internal and external formations.

The conclusive question of this paper concerns the consistency of the bio-informational compound from the point of view of aesthetics. What condition enables the *aesthetic* dimension of an assemblage created along the compositional vectors of a morphogenetic physics? How can the notions of information and image which I have worked out in terms of a pragmatics of Simondonian thought, be *synthesized* upon the rhyzomatic plane of consistency opened by the joint thought of Deleuze *and* Guattari?

In this regard it is useful to examine the lines of construction of Deleuze's concept of multiplicity. Anne Sauvagnargues points out that the French philosopher combines hereby the Simondonian operation of transductive relation with the Bergsonian concept of perception *qua* subtraction as developed in Matter and Memory. As a result of this combination, multiplicity does not (only) consist into an operation of $n + 1$, i.e. of a genetic addition of singularities. The aesthetic specificity gets instead grasped by the asymmetrical condition of perceiving the *subtracted* singularity $n - 1$. It is through the logics of this very operation, that Deleuze provides the operations dramatizing Simondon's idea of individuation—still framed into the "unitary pole" of onto-genesis—into the more complex realm of a theatre of *hetero*-genesis (Sauvagnargues 258). The field of forces opened by the multiplicity of the disparate allows the act of synthesis to become creative. In A Thousand Plateaus Deleuze and Guattari explain:

> The more rarefied the atmosphere, the more disparate elements you will find. Your synthesis of disparate elements will all be the stronger if you proceed with a sober gesture, an act of consistency, capture or extraction that works in a material that is no longer meager but prodigiously simplified, creatively limited, selected. (Deleuze and Guattari 1987, pp. 344–345)

It is this at the level of this *limited* synthetic realm between heterogeneities, that 'image' in its liminal state between sensible (external) dimensions of aesthetics and processing mechanisms of the brain, might become the generative condition for a different kind of creative formation. And activate novel form of potentialities implicit in the material substance of information itself.

## References

Chabot, P.: Philosophy of Simondon: Between Technology and Individuation. Bloomsbury, London (2013)

Berardi, F., Sarti, A.: RUN. Mimesis, Milano-Udine (2008)

Deleuze, G.: Difference and Repetition. Columbia University Press, Columbia (1995)

Deleuze, G.: Desert Islands and Other Texts 1953-1974. Semiotext(e), Cambridge (2004)

Deleuze, G., Guattari, F.: A Thousand Plateaus: Capitalism and Schizophrenia : Books. University of Minnesota Press, London (1987)

Guattari, F.: Cartographies Schizoanalytiques. Editions Galilée, Paris (1989)

Guattari, F.: Les Années d'Hiver: 1980–1985. Les Prairies Ordinaires, Paris (2009)

Mackenzie, A.: Transductions Bodies and Machines at Speed. Continuum, New York (2002)

Margairaz, S.: From Intuition to Transduction; beyond the heuristic value of analogy. In: A Comparative Reading of Simondon and Bergson. L'Harmattan, Paris (2010)

Moles, A.: Informationstheorie und ästhetische Wahrnehmung. M. DuMont Schauberg, Köln (1971)

Sauvagnargues, A.: Deleuze: l'Empirisme Transcendantal. Presses universitaires de France, Paris (2009)

Simondon, G.: Du mot d'Existence des Objets Techniques. Aubier, Paris (1958)

Simondon, G.: L'individu et sa Genèse Physico-Biologique. Presses Universitaires de France, Paris (1964)

Simondon, G.: The Genesis of the Individual. In: Crary, J., Kwinter, S. (eds.) Incorporations. Zone Books, Cambridge Mass (1992)

Simondon, G.: Imagination et Invention. Editions de la Transparence, Chatou (2007)

Toscano, A.: The Theatre of Production: Philosophy and Individuation between Kant and Deleuze (Renewing Philosophy). Palgrave Macmillan, Basingstroke (2006)

Toscano, A.: Gilbert Simondon. In: Jones, G., Roffe, J. (eds.) Deleuze's Philosophical Lineage. Edinburgh University Press, Edinburgh (2009)

# Part II
# Morphologies, Culture, and Spaces

# Chapter 5
# The Cultural Individuation of Human Language Capacity and the Morphogenesis of Basic Argument-Schemata

**Wolfgang Wildgen**

**Abstract** Language capacity unites all existent human populations. Therefore its morphogenetic field must have originated with the species Homo sapiens (ca. 200,000 y BP). We assume that at the origin this field was only poorly expressed (via some protolanguage), but unfolded with the rise of human cultures (technologies, art, myth and other symbolic media). This development occurred in the last 100,000 y. In this context, Thom's criticism of evolutionary theory in Darwinian terms can be reconsidered. He accepts the Darwinian principles but adds the role of morphogenesis and individuation. This article sketches the morphogenesis of basic sentence (argument) structures applying results of morphodynamics and the individuation of linguistic tools in phonemics, lexicon and grammar.

## 5.1 Darwin Revisited and the Relevance of Morphogenesis

In an article published in 1983,[1] Thom discussed Darwinism, its forces and faults. Its major force was also the major reason for its opponents against it. It rendered God as a basic explanatory force of evolution superfluous. As such, it continued the trend established by Condillac in the 18th century and announced the triumph of physical scientism at the end of the 19th century. Its second force was to introduce a fictive space of possible evolutions out of which selection could choose. This space became more concrete with modern genetics and the deciphering of the DNA. The weaknesses remain. First selection concerns the phenotype (the individual animal, its survival and the creation of progeny) and the transmission of the effect from the genotype is less direct than presumed (epigenetic and global bodily physiological

[1] Published first in the *Revista di Biologia*, 76 (1) and then again in Thom (1990: 599–605) and named: "Darwin cent ans après".

W. Wildgen (✉)
Institute of General and Applied Linguistics, University of Bremen, 330440,
Bremen 28334, Germany
e-mail: wildgen@uni-bremen.de



A. Sarti et al. (eds.), *Morphogenesis and Individuation*,
Lecture Notes in Morphogenesis, DOI 10.1007/978-3-319-05101-7_5

factors interfere). Second, the survival and expansion of new variants (based on unpredictable mutations) depends on many body-internal, social and ecological contexts. Third, some effects may be self-referential, insofar as new variants change the survival conditions. This can be the case *inside* the organisms if variants trigger structural reorganizations, or *outside* if moving and very active animals change their environment. This effect became clear with humans and most dramatically since the industrial revolution, which led to a new stage, the so-called *anthropocene*, in a world deeply changed by man (anthropos).

These critical remarks by Thom have consequences for evolutionary psychology, linguistics and the evolution of human culture, because in highly complex and large systems complicated processes of self-organization and self-referential effects occur. Simple Darwinian mechanisms are clearer at the level of molecular evolution (cf. Eigen and Schuster 1979).

The major Darwinian scenarios of language evolution are described in Wildgen (2004: Chap. 2). It became clear that at least two rather rapid and important changes occurred:

(a) The transition from ape-like behaviour in australopithecines (3–4 million years (my) before present (BP)) to Homo erectus/ergaster (2 my BP). It doubled cranial capacity, introduced tool use and led to larger social groups.
(b) The transition from descendants of Homo ergaster in Africa to Homo sapiens (300,000–200,000 y. BP). After a bottleneck around 120,000 y. BP, this species expanded in Africa and out-of-Africa (mainly after 70,000 y. BP). The changes were of a biological nature, i.e., they were fixed in the genetic heritage, although accompanied by cultural (technological) and social changes. It seems plausible to explain these changes applying Darwinian principles. In the case of language, this led first to the evolution of a proto-language with Homo erectus and second, with Homo sapiens, to the basic human language capacity shared by all humans. The time lapses are big enough for such an explanation:

   - From the last common ancestor (LCA) with chimpanzees (7/6 my BP) to the Australopithecus (4/3 my).
   - From the Australopithecus to Homo erectus/ergaster; i.e. from 4/3 to 2 my).
   - From the paleo-species Homo erectus[2] to the speciation of Homo sapiens (200,000 y. BP).

In the period since the first out-of-Africa dispersion of Homo sapiens (70,000 y. BP), the lapse of time is too small to explain deep biological changes along the explanatory lines of neo-Darwinism and a hypothesis, which places language origin after this date cannot explain the unity of human language capacity independent from the geographical distribution of mankind. Sophisticated cultural innovations appeared with the Neolithic revolution (beginning after the last ice age, i.e., approximately 10,000 y. BP), the first large and highly organized civilizations

[2] The last surviving non-human species was the late Neanderthal man in Europe (extinct between 37,000 and 30,000 y. BP), i.e. before the maximum of the last ice age (25,000–20,000 y. BP).

in Egypt and Mesopotamia, and the emergence of writing. It is not plausible that evolutionary processes affected basic human capacities in this period, because the Darwinian mechanisms of mutation and selection ask for larger time intervals due to mutation rates and necessary time spans for the distribution and dominance of selected mutations.

Rather than returning to religious explanations (or other Deus ex machina hypotheses like macro-mutations), we should consider the factors neglected by neo-Darwinism that Thom (and many others) have noted. The first is *morphogenesis* which is a background force directing and modifying Darwinian selection, the second is *individuation*, which operates in the transition between genotype (the biological or genetic type) and phenotype, the individual or social "body", through which the genotype is expressed and which survives, reproduces itself or disappears. In this move the Darwinian principles are not rejected, they rather define one set of basic forces.

### 5.1.1 Morpho- and Semiogenesis

In epigenetic processes and in embryology a structural framing occurs which puts a limit to new genetic variants and their bodily expression, i.e., they shape the space of possible evolutions. In the social, cultural and ecological domain other morphogenetic processes occur, which are more properly called semiogenetic, insofar as the perception and mental reaction to the environment are a requisite of form-giving, imitation. Cultural transmission replaces or goes in parallel with genetic transfer. Semiogenesis redefines the relevant environment, changes the selective forces and thus indirectly influences the genetic outfit and its epigenetic expression. Two highly relevant transitions after hominization had a major impact on the further development of human symbolic media and languages:

(a) The culture of painted caves in the late Palaeolithic (37,000–16,000 y. BP). A rich corpus of paintings, drawings, sculptures, and abstract (quasi-writing) symbols documents this period, which extended over Central, and Eastern Europe under the conditions of the last ice age.
(b) The new technologies of farming and cattle breeding, which led to many cultural innovations, e.g. writing and urbanization. It led to the first large-scale societies in Egypt and Mesopotamia (beginning around 5,000 y. BP).

### 5.1.2 The Impact of Individuation

The Darwinian hypothesis concerns primarily biological species (cf. the title of Darwin: 1859, *Origin of species*). However, the processes of survival and selection operate on the level of the individual. This is clearest in Darwin's favourite example: sexual selection. It occurs in a situation where both sexual partners meet. They either reject or accept one another and the transfer of genes is blocked or enabled. This scenario presupposes *individuation* and a *context* (time span, place and the presence of other agents like rivals). Individuation is, as Thom points out in his paper "Individuation et finalité" (Thom 1990: 206–217), defined by *topological closure*, *connectedness* and in many cases by its *contractility*.

Contrary to the individual plant or animal, the species is not closed and connected. However, a tribe or family group is connected via its biological origin (progeny, shared genes) and by its social/economical autonomy. It is therefore a limit-case of individuation and mediates between the more general level of species and the specific level of individuals.

Internally, an individual is defined by the natural limits of life (i.e. birth and death) and the forces, which sustain its life. If life is endangered, there exist appropriate dynamics to recover, to repair damages and to avoid death. In this perspective, individuation has an implicit finality: survival under danger and risk. Thom uses the metaphor of the crater (mathematically a potential; in Fig. 5.1 a parabola with negative gradient; the minimum is the place of rest). If the state remains inside in spite of perturbations and risks, it survives. If it is driven out of the survival hole, it dies (or it changes to a different mode if the attractor landscape is richer; cf. the crater at the right of Fig. 5.1).

A tribe or family group may disappear if its reproduction rate is too low. Thus, in small and separated groups, a scenario characteristic for Neanderthals, the loss of some younger females may drive the group to extinction; if this occurs often and becomes a trend, the species may be extinct after a number of generations if the overall population is small.

By these dynamics, individuation can drive the evolutionary process, i.e., avoid or allow extinction of a group or even a species. If individuals and groups of one species or subspecies share the ecology with others, their success may enhance the extinction of the other species, if mixture is genetically or behaviourally excluded or rare (dangerous).[3] Individuation is also the bridge for innovation either biological or cultural. For biological innovation, the individual, which shows new features due to some biological change,[4] must first survive in the environment of individuals without this change and then be successful in reproducing. These conditions decide if the new capacities of the individual survive and finally change the characteristics of a group or even a species. In the case of cultural innovation, it must be perceived

---

3 Thus, Neanderthal individuals seem to have mixed with individuals of the early Homo sapiens when they met around 100,000 y. BP, but mixture in Europe after 40,000 when Homo sapiens migrated massively into Europe was rare and did not prevent the extinction of the Neanderthals (possibly their offspring had health problems and rarely survived).

4 In a study on genetically identical mice Freund et al. (2013) showed that individual differences increase with age due in "cumulative roaming entropy" and "nonshared environment" (ibidem: 756). These behavioral differences have a neural correlate in the hippocampus and are thus of a biological nature. The question is if and how such differences may be genetically transmitted; they may have an impact on survival and reproduction.

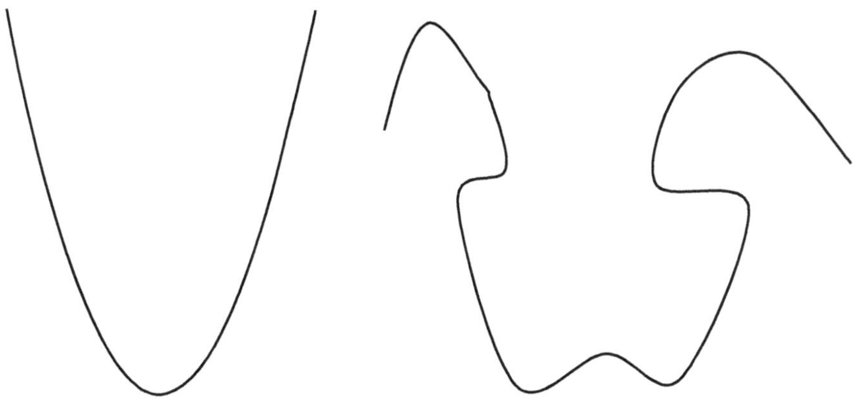


**Fig. 5.1** The parabola with negative gradient as schema of stable existence and a more complicated crater (cf. Thom 1990: 208)

as positive by other individuals in order to be imitated and to become part of a cultural heritage. In every case, the change must be perceived, evaluated such as to trigger a kind of new kind of behaviour, i.e., the effect relies on semiotic (sign-related) factors. The evolution of symbolic behaviour is just a special case of such a process of individuation and the semiotic socialization of individual innovations.

### *5.1.3 Tradigenetic and Ratiogenetic Processes*

Tembrock (2004) describes the self-organization of information in semiosis (of biological organizations). He distinguishes tradigenetic and ratiogenetic processes. In the first case socio-cultural values, standards, techniques are transmitted and conserved inside a social group (family, tribe, geographically connected set of tribes). This concerns techniques of chase and harvesting, tool industries, house building, the use of fire and cooking, etc. Innovation occurs more or less by chance, is imitated, further distributed and finally integrated into the heritage of a group. Ratiogenetic processes enhance such chance events by distinguishing individuals or professional groups for their directive function. These persons may be elder men or women, priests (shamans), elected chiefs, etc. They do not only incorporate the heritage of the group (in their life memory), but can plan and direct specific innovations or dramatic changes which have a deep influence on the fate of a population. Thus, the figure of Moses, who led the people of Israel out of Egypt, the prophets and founders of religion or significant statesmen (see Alexander, Augustus in antiquity) and their helpers (a small subpart of the society) can rationally move a given society into a specific direction. Since industrial revolution (England 18th century), the French revolution (end of the 18th century) and socialism (19th century) technological, scientific and political innovations have influenced the development of humanity. They had even an impact on the climatic and geological state of the earth (cf. the exploitation of resources deep in the underground and in the oceans and changes in the chemical structure of the atmosphere, atomic tests and wars). This development is sometimes called a new geological era, named *anthropocene*. In the end, the environment of man and thus the conditions for the survival of man could be altered by ratiogenetic processes (including the extinction of humanity in a self-made global catastrophe). Insofar as ratiogenetic processes refer to individual minds, i.e., the imagination and planning of individuals, they are part of individuation. In the process of transmission, the traces of this origin may disappear as many people contribute to its elaboration, transmission and the establishment of new standards. The importance of individuation becomes obvious in societies, which suppress individual creativity and pay this default with a slower growth of science, culture and economy.

## 5.2 The Morphogenesis of Linguistic (Symbolic) "Tools"

As Saussure (and more specifically) Martinet in the tradition of structuralism and structural functionalism (cf. Wildgen 2010; Chap. 3) have pointed out, human language shows characteristically two levels of "articulation":

- The level of phonic classification (phonemics) and architectures (phonotactics).
- The level of meaning-related classification (lexicology) and combinatorics (morphology and syntax).

The base line is the level of bodily signals expressed by movements (of the tongue, the hand or other body parts), which is encountered in many animals and used for communication. The specificity of humans is their preference for the phonic channel and the enhancement of phonic capacities in terms of speed in articulation, precise control of articulatory targets and high auditive resolution. We assume that since the australopithecines auditory capacities were enhanced due to their endangered life in the savannah (cf. Calvin and Bickerton 2000, p. 110). The articulatory capacities beyond the level of our last common ancestor with chimpanzees (LCA) were probably gradual and exploited the higher level of auditory control. At least in the beginning this was a spandrel effect. With the increase of a basic lexicon of calls and the necessity to distinguish them with high precision the phonic capacities increased gradually. Relevant stages of this unfolding could have been:

(1) Basic types of articulations (beyond mastication and ritualised smacking) as found in chimpanzee vocalizations and basic calls with semantic distinction (3–4 types of referential calls related to types of predators). At this level, the feature of *double articulation* is already present.
(2) Higher auditory resolution in australopithecines allowed for the increase in call diversity (e.g., beyond 10 calls with beginning morphological structures via

self-organization of this repertoire). Calls with a clear phonetic distinction between vowels and consonants (CVCV-sequences) and words with two syllabic kernels could have evolved at this stage. This may be called a morphogenesis insofar as the space of vowels is separated into subfields (front-back, high-low) and the syllabic structure of words emerges.

(3) In the stage of Homo erectus, tools were produced controlled by specific standards (an industry of tools appeared) and we may assume the cultural unfolding for linguistic "tools", used in the teaching of lithic and other technologies. Such a standardizations means that:

- Phonemic types are fixed.
- Syllabic morphemes with a variety of different forms become standard. They enable a lexicon beyond 10 (possibly up to 50 calls or utterances).

This stage may be called a *protolanguage* (cf. Bickerton 2000, and Wildgen 2004: Chap. 8).

(4) If at the phonic level a kind of syntagmatic structure existed (forced by a larger lexicon and its natural self-organisation), the semantic organisation may have been either holophrastic or at the level of two-word-utterances in early child-language (with one pivot term as centre and a variety of occasional second terms). In Wildgen (2004: 168) I have formulated the (first) restriction of valence complexity, i.e., utterance were either holophrastic (value 0) or had maximally one additional (facultative) argument (valence 1) as in child language or comparable to intransitive sentences in actual adult languages.
In the unfolding of a larger lexicon, the relevant and recurrent categories of the perceived and practically used flora and fauna and kinship relations were phonetically coded. This morphogenesis differentiated the space of *pregnancies* (cf. Thom's term) or *affordances* (in Gibson's terms).[5] The relevant life-space is cut into categories in the same way as the auditory and articulation space was cut into subfields in phonic audition and articulation (cf. Petitot 1985; Wildgen 1998). In this line, the discrete and highly combinatory character of human language emerges.

(5) Finally, in human speciation, i.e., due to the behavioural and biological separation of the branches of Homo erectus, e.g., Homo neanderthalensis from Homo sapiens around 200,000 y. BP, the language capacity of our species was finally developed.[6] Beyond a higher phonic competence (precision and speed, longer utterances), the critical transition concerns mainly valence patterns, i.e.,

[5] Cf. The contributions in Wildgen and Plümacher (2009; in German), and Wildgen and Brandt (2010: English and French) to this topic.For the term "affordance' see Gibson (1966).

[6] Currently a comparable although different language capacity is assumed for Homo neanderthalensis. It allowed for a restricted communication with Homo sapiens populations if they met in a cooperative context. Current genetic analyses showed traces of Neanderthal genes in the populations, which had left Africa, probably introduced around 100,000 y. BP into the genome of Homo sapiens.

the consolidation of the bipolar sentence structure (case-frames with one argument: subject–predicate or ergative–predicate constructions; cf. Wildgen 2002). The argument complexity was further increased by the introduction of secondary arguments leading to sentences with valence two and three (e.g. transitive and bitransitive constructions). The facts that statistically valence-one patterns dominate, followed at some distance by valence two-patterns is a trace of this evolutionary sequence. Some languages still prefer very primitive patterns (cf. the "minimal language thesis" by Gil 2006). The relevant steps were:

- the bifurcation from holophrastic utterances to bipolar, dual ones,
- the expansion of argument patterns and predicate typology beyond the basic (valence one) utterance (with a subject or ergative case). The complexity of the larger frames mirrors the increase in cultural complexity, which had to be managed in cognition and sign-behaviour.

The morphogenetic unfolding of human language capacity follows from the germ already present in animal communication at the level of the LCA and in the context of an enlarged brain with greater neural capacities in the fields of (auditory) perception, visual categorization of environmental cues (affordances) and memory. At the stage of Homo erectus and its followers (Homo heidelbergensis and others) the neural demands of phonetic communication became a selective features fostering brain size increases, although initially brain size may have rather be enhanced by new motor, perceptual and probably cognitive demands in larger social groups (cf. Dunbar 2002). Thus selective processes cooperate with morphogenetic unfolding, insofar as they stabilize the rise and self-organization in the morphogenetic field.

## 5.3 Thom's Conjecture and the Morphogenesis of a Complex Language

In the following, I shall put into parallel two types of "unfolding" of morphogenetic germs, with the aim of explaining the second via the first one[7]:

1. The unfolding of process schemata as mirrored in the hierarchy of elementary catastrophes leads to "semantic archetypes" proposed by Thom and elaborated in Wildgen (1982, 1985). These archetypes are semantic in the sense that they abstract real processes to a schema, which is, then at the basis of cognitive and linguistic gestalts.

[7] Cf. Wildgen (2004) for a set of principles governing this process. For the sketch of an "evolutionary grammar' see Wildgen (2009).

2. The unfolding of symbolic and linguistic behaviour at the stage of early Homo sapiens (or his immediate predecessors) mirrors a similar emergence of schematic complexity, i.e. complexity in the world is projected into complexity in language (and linguistic cognition). Semiotically, this operation is indexical, because the dynamics in the world are at the basis of the dynamics in language, and iconic because the schemata define a similarity between the world (objects, processes, features) and the symbolic or linguistic medium. This very abstract motivation of linguistic structures is the reason for their efficiency in human cognition and behaviour and therefore has a high selective advantage, which drove its emergence, stabilization and further elaboration.

## *5.3.1 Sketch of the Topologico-Dynamic Basis (Thom's Conjecture)*

Ad (1): The basic idea in Thom's semiotic writings was to link verb-valences and the classification theorem of catastrophe theory. Tesnière (1959) considered the maximal valence of verbs to be three (with a fourth 'actant' in the causative construction). The basic "morphologies" furnished by elementary catastrophe theory (cf. Wildgen 1982) can be interpreted as a universal set of perceived or enacted situational schemata, which are exploited by human languages as minimal scenarios for utterances. If we consider linear paths in an unfolding, i.e. in the phase spaces sketched in Fig. 5.2, we can classify a basic set of processes (cf. Wildgen 1982 and 1985 for the elaboration and critical analysis of the set proposed by Thom).

On this basis, René Thom formulated his conjecture (cf. the Chaps. 10–13 in Thom 1983; the basic set had been derived in 1972):

### 5.3.1.1 Thom's Conjecture

Given a dynamic situation, the analysis of structural stability cuts out pieces of the continuous process:

a. in the neighbourhood of singularities (catastrophes),
b. these segments have a maximum complexity of 3 (with one component) or 4 (with two components).

If we consider Thom's conjecture and the list of process-schemata derived from elementary catastrophes, we come to the following conclusions:

- A finite (small) list of formal process scenarios is derived by considering states, continuous processes, and transitions/changes along linear paths in elementary catastrophes.

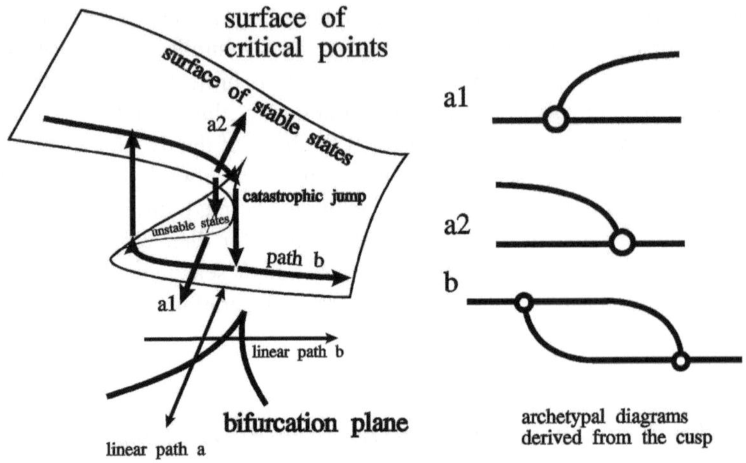


**Fig. 5.2** The derivation of archetypal diagrams from the "cusp"

- The static stable points, lines and surfaces are interpreted cognitively as mental attractors and linguistically as nominal entities, specifically nominal roles in minimal sentences. The stable process types of events are interpreted cognitively as mental scenarios and linguistically as predicative centres of minimal sentences.

The lexicon of verbs with its valence patterns and selectional restrictions is in many languages a very systematically organized field, and the starting point of every model of the sentence is the main verb (or the predicative centre) of the sentence. If the basic problem of verb valence, i.e. of the gestalt-patterns (constructions) represented by verbs and their arguments is solved, then the question of how the meaning of sentences is organized in minimal utterances can also be solved.

Ad (2): Predecessors of Homo sapiens, who lived in Eastern and/or Southern Africa some 400,000–300,000 years before present (BP) belonged to a branch of the paleo-species Homo erectus (the African branch is mostly called Homo ergaster). Well-known branches of those subspecies that left Africa since 700,000 y BP are Homo heidelbergensis (a skeleton was found near Heidelberg and is dated around 400,000 y. BP) and Homo neanderthalensis (this species probably met the Homo sapiens population in the Near East and exchanged genes with it. Later both species lived in Europe and the Neanderthal man died out some 30,000 y. BP). In this phase of technological and cultural evolution, the Levalloisian technique of flint shaping was invented and perfected around 100,000 y. BP; around 64,000 y BP the use of bow and arrow was invented, painted caves appeared after 40,000 y BP. We assume that language was shaped in the period of human speciation (300,000–200,000 y. BP) and elaborated before the out-of-Africa move (beginning 70,000 y. BP). The only systematic and law based explanation exposed up to now (beyond simple narratives about possible origins) refers to a topologico-dynamic measure of schema complexity (cf. Wildgen 1982, 1985, 1994 for the basics and its application to the evolution of language in Wildgen 2004).

In order to define such a complexity measure, I shall use the basic catastrophe theoretical analysis shown above and interpret it in the light of evolutionary theory. I shall take grasping (with the hand) as the starting point (cf. for the semiotics and linguistics of the concept *hand* Wildgen 1999).

### *5.3.2 Grasping as Basic Intentional Cognitive Schema Semantically Coded in Language*

The action-concept GRASP involves two stable entities: the body (the hand) and the object. Every point on the lines in Fig. 5.3 is an attractor. The schema is called the GRASP-schema and may be translated into sentences like:The man catches/ grasps the rabbit/ stone

Predecessors of humans (e.g., Homo habilis) already had a hand with the opposition of thumb and fingers, but some features are still linked to climbing (as in gorillas and chimpanzees). The Homo erectus had a hand, which was adapted to strong grasping (as places on the bones, where muscles are attached, show). This was still the case for Neanderthals. The distinction between several types of grips leads to more elaborated schemata as soon as more precise manipulations on objects and instruments are needed. This concerns mainly manner-distinctions, which may surface in the lexicon of verbs or in specific adverbial attributes.[8] The fact that a cognitive bivalent schema and a manner component can be cognized does not necessarily mean that it could be transformed into phonic signals. If we assume a frequent vocalization (inferred from the evolution of the sub-laryngeal tract) and a steady increase of memory (due to the growth of the brain) linked to an advance in social cognition, it becomes clear that this cognitive schema and subsequent ones are pre-adaptations for the evolution of verbal phrases or valence patterns in sentences. Thus, in order to represent important and recurrent actions verbally the cognitive schema of grasping could be used as a kind of ground for iconic/metaphorical transfer to all kinds of manipulations on objects. As soon as instruments were used, this schema could be iterated.

---

[8] We start with rather complex schemata. Naturally a schema of frontier and the transition of frontier lines preceded the grasp schema, see for a short summary of catastrophe theoretical semantics Wildgen (2005). It may already be observed in the behaviour of chimpanzees who control their territory and attack other chimpanzees near the frontier line in order to prevent their intruding. The semiotic relevance of this schema was discussed (in relation to evolution) in an article by Per Aage Brandt: *The Principle of the Wall—Or How Space (may have) Shaped Semiosis. A Note (download from*: http://papers.ssrn.com/sol3/papers.cfm?abstract_id=2304745).

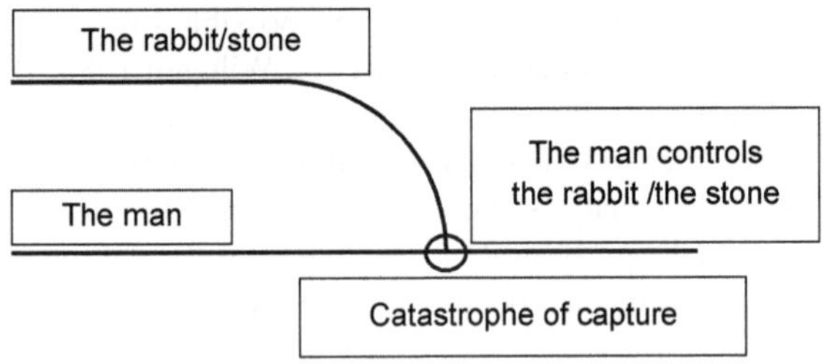


**Fig. 5.3** Catastrophe schema of GRASP

(a) The man/women (A) takes a hand-axe (B) to move/change/kill … object (C).
(b) The man/women (A) takes a stone/bone ($B_1$) to hit/shape the pebble ($B_2$) which should later kill the animal (C).

### *5.3.3 Beyond the Grasp Scenario*

The manufacturing of stone tools (and *a fortiori* of tools shaped with the help of stone tools) goes cognitively beyond the grasp scenario (cf. Wildgen 2004: Chap. 4). One hand (or one foot) must fix the pebble, the other hand grasps the stone or bone which hits the stone. Finally the planned breaking off subtracts material from the chosen stone and after several strokes the desired sharp edge of the pebble is produced (cf. the Levalloisian technique). This scenario involves two objects, two hands and a change in shape of the pebble (the separation of parts from it). René Thom (1983: 182) proposed the *excision* schema, which is presented in a modified version in Fig. 5.4.

This schema contains four symmetric "grasping/emitting" sub-schemata (simple instrumental action) and one further "emitting" schema (dark circle; the extracted part). The first four are integrated in the (double) transfer-schema. The integration of the shaping by the tool is on a higher level of complexity (it has corang 2; cf. Wildgen 1985: 201). In fact, a linguistic description of the action normally requires more than one basic sentence pattern in actual languages.

### *5.3.4 The Configurational Structure of "Giving"*

Beyond the field of instrumental action and tool production, which was the major centre of collectively standardized, rule governed behaviour and as such a model for the linguistic tool featuring as sentence/proposition in actual languages, we must consider schemata which organize social action and its equilibration. The exchange of goods, presents, the distribution of food etc. was at the origin of human societies and cultures. Marcel Mauss based his theory of society on the relevance of presents and counter-presents (without money or equivalents) and called them "total social

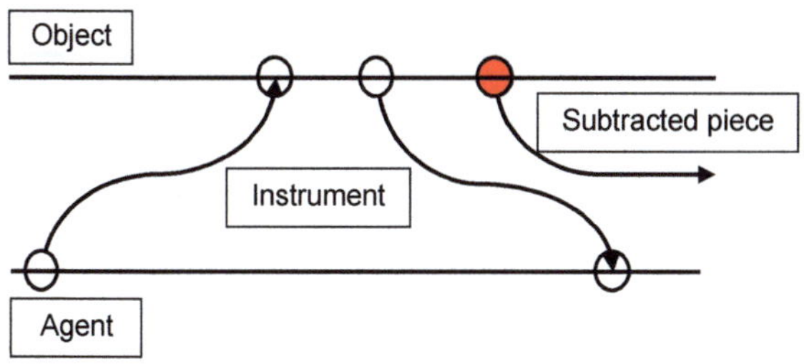


**Fig. 5.4** Schema of shaping an object with an instrument via excision

facts" (cf. Mauss 1968). Saussure categorized language ("langue") as a social fact in a similar spirit. These reflections point at a social origin of linguistic tools (grammar, lexicon, phonetics) beyond a behavioural one.

The basic schema or prototype of "giving" can be configurationally described by the structure of basins, where each basin represents the specific positions of sender, receiver and object. At the beginning and at the end of the series one observes two basins (i.e. *attractors/minima* in catastrophe theory); at the centre of the series, a third attractor appears, grows and finally disappears. If temporal deformations of contour diffusion are considered, bifurcating events can be discovered and categorized (cf. for further details Petitot 2011, p. 270).

The intermediate, symmetrical scene is the most unstable one. Both agents concentrate their control on one target, and their control must be coordinated in order to secure the smooth exchange. Thus, if A releases his control before B takes the object, or if A holds the object tight, although B seizes it, the character of the process is dramatically changed and degenerates to "A loses, drops the object" or "A and B compete for the object C". Thus, the unstable state of exchange is the "junction" of the process, the point of maximum coordination of the controls. In Fig. 5.5 five major phases are distinguished applying the catastrophic schemata called "EMISSION", "CAPTURE" and "TRANSFER" (transition) between HAVE1 and HAVE2. The phases can be further subdivided in view of the dominant perspectives (from M1 or from M2). The line of TRANSFER separates HAVE and HAVE NOT for M1 and M2.

In English, we find the following subclasses for 'give':

| | |
|---|---|
| *receive, take, take off, rob, steal*: | CAPTURE |
| *give, donate, exchange*: | TRANSFER (implying EMISSION and CAPTURE) |
| *buy, buy from, purchase, shop*: | TRANSFER + CAPTURE (foreground) |
| *sell, lend/borrow, return*: | TRANSFER + EMISSION (foreground) |

The concept of dominance in a dynamical system can also be used for the modelling of topicalization and passive transformation (cf. Wildgen 1990). The exchange of goods implies a specific force, which is more symbolic than the objects exchanged, i.e., the *value system* which allows for a satisfying exchange. This system may be qualitative, e.g. there are generalized goods (gold, silver, cupper,

**Fig. 5.5** The phases of the TRANSFER schema

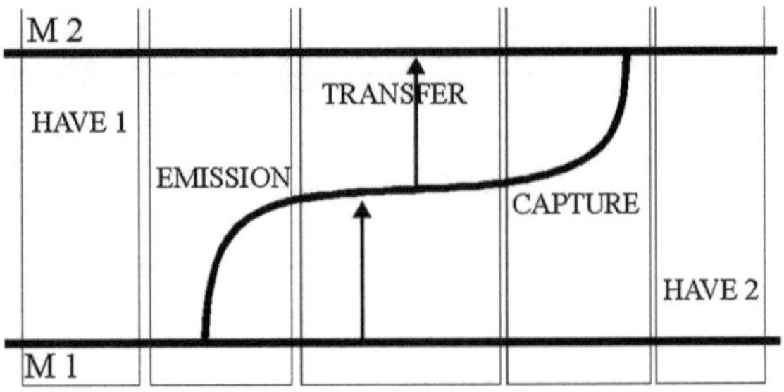


salt, feathers etc.) which establish a tertium comparationis (a neutral medium, a catalyst) for exchanged goods, or quantitative as measures for the amount (numbers), the weight or the size of exchanged objects and materials.

## 5.4 Individuation of Language Capacities

In order to learn a language children need a proper linguistic environment. As Kaspar Hauser phenomena show, there is a critical zone for this individuation. If the child does not acquire basic linguistic skills in the early period (up to 4 years) and elaborate it until puberty (11–15 years old), a proper skill can only be acquired under severe restrictions. In a self-referential loop, the acquisition of linguistic skills and their elaboration (e.g., by linguistic art like singing, poetry, narration, argumentation) set the context for the next generation of speakers and their communicational achievements and standards. In a different direction, linguistic skills may also be deteriorated if the transition of skills between parents (in appropriate social contexts) and children is very poor or even disrupted. This was the case in historical slavery, where young slaves were separated from their families and grew-up in a multilingual and linguistically very poor environment (selection camps in Africa, slavery ships and artificially disrupted social groups in the context of plantations). The pidgins and creoles, which resulted, show not only the loss of original linguistic patterns (mostly West-African idioms) and dramatic simplifications in the acquisition of the colonial language (Portuguese, Spanish, English, French, etc.); they demonstrate also that the underlying species typical language capacity can recreate a fully functional language (a creole-language) even after such dramatic losses.[9] This means, that individuals in deteriorated and disordered linguistic contexts are able to "invent" human language again and again. Human language is,

[9] In Bickerton (1984) a *language bioprogram* is proposed in the spirit of Chomsky's hypothesis of some inborn grammar ("universal grammar"). We reject such mechanistic devices and assume that the biologically fixed capacities are rather cognitive and behavioural than linguistic (in a narrow sense implied by the term *grammar*). Therefore, a topologico-dynamic basis is more plausible than low dimensional (linear) algebraic models (generative grammars) discussed by Chomsky.

therefore, more that a tradition or a chance invention. Human societies can under minimal conditions of social contact and information exchange, compensate the loss of a linguistic tradition and initiate new traditions which gain stability and after a period of consolidation can even fully replace the languages lost. Thus, many creole-languages developed in the 17th century are still used and even became national languages in some communities. The possibility of individuation in the morphogenetic field of linguistic unfolding is therefore the major stability factor in human languages.

## 5.5 Tradigenetic and Ratiogenetic Processes

In the further historical development, human languages were not only diversified due to migration and the separation of linguistic groups (a contrary move of globalisation may be observed in the last centuries), they have also modified their functional profile. This was due to the introduction of writing and institutional instruction. This further unfolding of human symbolic and communicative skills may be called ratiogenetic, insofar as artificial linguistic tools: writing systems, languages of science and other special purposes, mathematics, computer languages are designed which secondarily change everyday communication. This topic goes beyond the aims of this article and concerns the future of human languages. Basically, they are still unfolding the biological potential created with human specification, i.e., some 200,000 years ago in Africa.

In comparison with Darwinian theory morphogenesis and ratiogenetic processes reintroduce the aspect of finality which had been cancelled in order to ret rid of creationist "explanations" (theological or more generally *Deus ex machina* explanations). The new type of finality is due to the rise of civilizations and man's will to shape his own future. René Thom says in a similar context (cf. Thom 1990: 217):

> L'élaboration des moyens en vue de la réalisation d'une fin dépend (en technique et en biologie) d'une situation "culturelle" préexistante, et d'une certaine stabilité des conditions extérieures.[10]

Nevertheless, morphogenetic processes do not lose their significance in this enlarged context for two reasons:

(a) Morphogenesis defines already a space of possible evolutions from which cultural processes and rational design may choose under specific conditions. Chance factors have their place in this context, but morphogenetic gradients are more significant, insofar as they define the frame for selection of random changes.

[10] Translation by the author: "The elaboration of means in view of the realization of some goal depends (in technological and biological contexts) on a "cultural" situation existing before, and on a certain amount of stability of the external conditions."

(b) All ratiogenetic processes just concern adaptations, small modifications in the evolutionary space defined by morphogenesis and long terms Darwinian selection and chance processes (mutations in genetics and environmental chance processes due to many uncontrolled forces).

## 5.6 Concluding Remarks

In the analysis only a small area has been mathematically accessed, i.e., the morphogenesis of valence patterns in language. The process of individuation was discussed in Thom (1990, p. 207) in view of its topology. He mentions three major aspects:

(a) *Connectivity* of an individualized system. Currently this aspect may be elaborated in view of the underlying brain dynamics, i.e., in the field of neurodynamics applying stochastic dynamical system theory (cf. synergetics e.g. Haken 1996).
(b) *Contractibility.* Topologically individuals have the shape of a ball ("boule" in Thom's terms). Thus social hierarchies and religious systems are contractible, e.g., to a (temporary) king or a main God (as in monotheism).
(c) *Irreducibility.* The individual has a character, a soul, is a "persona" and cannot be separated into several centres without loosing its stability/identity (cf. schizophrenia).

The other major field of determining forces called *culture* (society) has also a tendency towards individuation, although its connectivity is different from that observable in individuals. Thom (ibidem) mentions the example of a table cover, with knife, fork, spoon, and shallow plate, soup-plate, cup, glasses. This configuration is topologically not connected, but the functional relations, practical use and the regular re-arrangement at the beginning of each meal establish a specific kind of functional connectivity and thus of *functional individuation*. Economical enterprises (cf. a factory) and administrations are comparable to this prototype. In the sequel, economic, administrative and political entities approach the individuality of human being. This intuition has already been mentioned in 17th century political philosophy, i.e., in the figure of the Leviathan by Hobbes. The frontispiece of his book shows the state (in his conservative version, the king) as a supra-individual whose body is made up by many human beings (see Fig. 5.6).

A proper mathematical treatment of political systems is beyond the aims of this paper, buts its feasibility seems plausible. In a similar mode, notions like *nation, religion, language* (systems) are supra-human entities, showing features of individuation. In biology animal states e.g. of ants, termites, bees show the transition from an individuality incorporated by single beings to the individuality of the

**Fig. 5.6** The picture of the state as a superhuman being: Hobbes' "Leviathan"

"state" (at the expense of the single constituents of these "states"). As the fate of many totalitarian regimes in the 20th and 21st century has shown, this is not the model human beings accept voluntarily and at long term.[11]

## References

Bickerton, D.: The language bioprogram hypothesis and second language acquisition. In: Rutherford, W.E. (ed.) Language Universals and Second Language Acquisition, pp. 141–161. Benjamins, Amsterdam (1984)

Bickerton, D.: How protolanguage became language. In: Knight, C., Michael S.-K., James, R.H. (eds.) The Evolutionary Emergence of Language. Social Functions and the Origins of Linguistic Form, pp. 264–284. Cambridge U.P., Cambridge (2000)

Calvin, W.H., Bickerton, D.: Lingua es Machina. Reconciling Darwin and Chomsky with the Human Brain. MIT-Press, Cambridge (2000)

Dunbar, R.: Human Evolutionary Psychology. Palgrave-Mac Millan, Princeton (2002)

Eigen, M., Schuster, P.: The Hypercycle. A Principle of Natural Self-Organization. Springer, Berlin (1979)

Freund, J., Brandmaier, A., Lewejohann, L. et al.: Emergence of individuality in genetically identical mice. Science **340**(6133), 756–759 (2013)

Gibson, J.J.: The Senses as Perceptual Systems. Houghten Mifflin, Boston (1966)

[11] The "dictator" in ancient Rome was only accredited his power for a restricted time in situations of extreme endangerment. In the movie *Leviathan directed by* Andrey Zvyagintsev and presented at the Cannes festival in 2014, Vladimir Putin is indirectly associated with the figure of Leviathan.

Gil, D: Early human language was isolating-monocategorical-associational, cf (2006). http://www3.isrl.uiuc.edu/~junwang4/langev/localcopy/pdf/gil06evolang.pdf
Haken, H.: Principles of Brain Functioning. A Synergetic Approach to Brain Activity, Behavior and Cognition. Springer, Berlin (1996)
Mauss, M.: Essai sur le don: Forme et Raison de L'échange dans les Sociétés Archaïques in Sociologie et Anthropologie. PUF, Paris (1968)
Petitot-Cocorda, J.: Les catastrophes de la parole. De R. Jakobson a René Thom, Maloine, Paris (1985)
Petitot, J.: Cognitive Morphodynamics. Dynamical Morphological Models of Constituency in Perception and Syntax. Lang, Bern, New York (2011)
Tembrock, G.: Ökosemiose. In: Posner, R., Robering, K., Sebeok, T.A. (eds.) A Handbook on the Sign-Theoretic Foundations of Nature and Culture, pp. 571–591. de Gruyter, Berlin (2004)
Tesnière, L.: Éléments de Syntaxe Structurale. Klincksieck, Paris (1959)
Thom, R.: Stabilité Structurelle et Morphogenèse. Interéditions, Paris (1972)
Thom, R.: Mathematical Models of Morphogenesis. Horwood (Wiley), New York (1983)
Thom, R.: Apologie du Logos. Hachette, Paris (1990)
Wildgen, W.: Catastrophe Theoretical Semantics. An Elaboration and Application of René Thom's Theory. Benjamins, Amsterdam (1982)
Wildgen, W.: Archetypensemantik. Grundlagen für eine dynamische Semantik auf der Basis der Katastrophentheorie. Narr, Tübingen (1985)
Wildgen, W.: Konfiguration und Perspektive im elementaren Satz: ein dynamischer Modellansatz. In: Bahner W. (ed.) Proceedings of the Fourteenth International Congress of Linguists, Berlin, 1987, pp. 1122–1125. Akademie Verlag, Berlin (1990)
Wildgen, W.: Process, Image, and Meaning. A Realistic Model of the Meanings of Sentences and Narrative Texts, Series: Pragmatics and Beyond, New Series, No. 31. Benjamins, Amsterdam (1994)
Wildgen, W.: Selbstorganisationsprozesse in der Phonologie. In: Wagner, K., Wolfgang W. (Hg.) Studien zur Phonologie, Grammatik, Sprachphilosophie und Semiotik, Nr. 6, pp. 123–137. BLIcK, Universitätsbuchhandlung, Reihe, Bremen
Wildgen, W.: Hand und Auge. Eine Studie zur Repräsentation und Selbstrepräsentation (kognitive und semantische Aspekte). Publications of the Center: Philosophical Foundations of Science, vol. 21, Univerversity Publications, Bremen. [Hand and Eye. A Study of Representation and Self-Representation (cognitive and semantic aspects). http://www.fb10.uni-bremen.de/homepages/wildgen/pdf/handundauge.pdf
Wildgen, Wolfgang: Dynamical models of predication. Sprachtypologie und Universalienforschung **55**(4), 403–420 (2002)
Wildgen, W.: The Evolution of Human Languages. Scenarios, Principles, and Cultural Dynamics. Benjamins, Amsterdam (2004)
Wildgen, W.: Catastrophe theoretical models in semantics. In: Köhler, R., Altmann, G., Pietrowski, R.G. (eds.) Quantitative Linguistics/Quantitative Linguistik. An International Handbook, pp. 410–423. de Gruyter, Berlin (2005)
Wildgen, Wolfgang: Sketch of an evolutionary grammar based on comparative biolinguistics. In: Röska-Hardy, L.S., Neumann-Held, E.M. (eds.) Learning from Animals? Examining the Nature of Human Uniqueness, pp. 45–59. Psychology Press, Hove (2009)
Wildgen, W.: Die Sprachwissenschaft des 20. Jh.s: Versuch einer Bilanz. de Gruyter, Berlin (2010)
Wildgen, W., Plümacher, M. (eds.): Prägnanter Inhalt—Prägnante Form. Thematic Volume of Zeitschrift für Semiotik, 31 (1–2). Tübingen: Stauffenburg (2010)
Wildgen, Wolfgang, Brandt, Per Aage (eds.): Semiosis and Catastrophes. René Thom's Semiotic Heritage. Lang, Bern (2010)

# Chapter 6
# Through the Looking-Map: Mapping as a Milieu of Individuation

**Mario Neve**

**Abstract** Mapping is not merely making maps. As demonstrated by a rich literature, mapping is an ancient cognitive activity, and its complexity as a process explains how troubling has always been for scholars to define it starting from the map as an object: indeed, mapping is quite an *absolute metaphor* in Hans Blumenberg's terms. Its main role has concerned (and still concerns) the production and conveying of *spatial information*. Spatial information is one of the most vital kinds of information for gathering and keeping human groups together, as far as relationships in their widest sense are involved. The present essay aims at showing how mapping works in setting milieus of individuation (following the insightful work of Simondon). In drawing mainly on the Simondon's concept of *invention* (along with the most recent literature on mapping), it is possible to show that the mapping is the leading information infrastructure at the root of the production of spatial information, that this feature has been made possible by the *chimeric* nature of mapping (combining logic and analogy), and that the radical novelty of the Geo-Web is based on the would-be solution to a problem already existing in the earliest examples of mapping.

## 6.1 Shaping a Place: Place and Individuation

Places and maps seem self-evident things. Everybody knows (or suppose to know) what they are, though providing, if requested, as many different descriptions as there are respondents.

M. Neve (✉)
University of Bologna—Ravenna Campus, Department of Cultural Heritage,
1, via degli Ariani, 48121 Ravenna, RA, Italy
e-mail: mario.neve@unibo.it
URL: http://www.unibo.it/faculty/mario.neve; http://www.researchgate.net/profile/Mario_Neve/; http://unibo.academia.edu/MarioNeve; http://www.mendeley.com/profiles/mario-neve/; http://goo.gl/pzM0P

 
A. Sarti et al. (eds.), *Morphogenesis and Individuation*,
Lecture Notes in Morphogenesis, DOI 10.1007/978-3-319-05101-7_6

As Robert Musil summed up in 1927 with his usual cleverness,

> Anything that endures over time sacrifices its ability to make an impression. Anything that constitutes the walls of our life, the backdrop of our consciousness, so to speak, forfeits its capacity to play a role in that consciousness (Musil 1978: 62).

Musil was speaking of monuments, then of that kind of landmarks we usually associate to urban places as one of their distinctive signs. The network of places and paths linking them are customarily considered inherent in the sense of belonging to a determined community as well. And, for this reason, they are taken for granted by insiders, because of their overfamiliarity, which makes them, in a sense, invisible.

It is such places' feature—their *naturalisation* (Iacono 2000) by insiders—which gives them the capacity to work out as primary (mostly unaware) elements of *orientation* (Almeida 2004), both individual (actions, choices), and transindividual (culture), to use Simondon's terms.

The efficacy of such naturalisation is guaranteed by an old evolutionary device, allowing to embody spatial information in the form of *spatial referential frames* (Tversky 2005), which Yi-Fu Tuan sketched *en géographe*:

> The mind learns to grapple with spatial relations long after the body has mastered them in performance. But the mind, once on its exploratory path, creates large and complex spatial schemata that exceed by far what an individual can en-compass through direct experience. With the help of the mind, human spatial ability (though not agility) rises above that of all other species. Spatial ability becomes spatial knowledge when movements and changes of location can be envisaged. Walking is a skill, but if I can "see" myself walking and if I can hold that picture in mind so that I can analyze how I move and what path I am following, then I also have knowledge. That knowledge is transferable to another person through explicit instruction in words, with diagrams, and in general by showing how complex motion consists of parts that can be analyzed or imitated (Tuan 1977: 67–68).

On the other hand, the relevance of places' 'invisibility', that is, of the function performed by places in the process of individuation thanks to their naturalisation, is even more underlined by the time's role, according to the two main ways spatial referential frames have been represented in most cultures: the *egocentric* way (spatial information organised *from* organism's position) and the *allocentric* one (spatial information *as independent of* organism's position) (Berthoz 2005).

Indeed, places are full of time. They show, if we pay attention to them, their different "time layers" (*Zeitschichten*)—to use Reinhart Koselleck's metaphor (Koselleck 2002)—that let them making sense for us, as a *concretisation* (Simondon 1958) of people, activities, relations, matter, ideas, words, images, and so on.

The physical places we live in every day make sense for us because we feel them through an inextricable stuff made of sensations, thoughts, memories (unconscious mostly), revealing their concreteness. According to the definition given by French geographer Augustin Berque, places are concrete because

> *concretus*, in Latin, it was the past participle of *concrescere*: to grow together. Actually [...] people, words, and things have grown up together; they have a shared story (Berque 2000: 18–19).

For such a reason places can be interpreted as a kind of collective, social memory of a group or community.

The bounded places of our everyday life—the streets of our town, workplaces, home, squares and monuments, and so on—are naturalised by cultural habits, routines, meanings acquired by education or social relations. Our feelings for places, our knowledge itself of them, are a mat weaving both adopted pasts and personal experiences. They belong to the dialectics between individual and society Leroi-Gourhan keenly outlined almost fifty years ago:

> Individuals at birth are faced with a body of traditions that belong to their ethnic group; a dialogue takes place, from childhood, between the individual and the social organism. Tradition is as biologically indispensable to the human species as genetic conditioning is to insect societies. Ethnic survival relies on routine; the dialogue taking place produces a balance between routine and progress, routine symbolizing the capital required for the group's survival and progress the input of individual innovations toward a better survival (Leroi-Gourhan 1965: 228).

Historically, in this perspective, modern age marked the beginning of a new experience of places. As transport and communication systems gradually evolved, the relationship between places and knowledge changed.

> In the seventeenth and eighteenth centuries, the acquisition of knowledge (of all but some members of the upper and middle class) was still dependent upon where a person was born (and subsequently lived out her/his life) and upon the corresponding local availability of schools, booksellers and other means of dissemination of empirical knowledge.[1] The nineteenth and twentieth centuries, in contrast, can be interpreted as a period of rapid homogenisation in the degree of spatial variation of availability of empirical knowledge (Thrift 1996: 112–113).

This means that, for a long time, social situations and places were considered closely connected, and that imagination often was the one and only way to escape, as it were, the constraints of places' boundaries.

> The relationship between physical place and social situation still seems so natural that we continue to confuse physical places with the behaviors that go on in them (…) Before electronic media (…) places defined most social information-systems. A given place-situation was spatially and temporally removed from other place-situations. It took time to travel from situation to situation, and distance was a measure of social insulation and isolation (…) Communication and travel were once synonymous. Our country's communication channels were once roads, waterways, and railroads. Communication speed was limited to the speed of human travel (…) A place defined a distinct situation because its boundaries limited perception and interaction (Meyrowitz 1985: 116).

[1] With this term Thrift means the socially structured knowledge in which grounds social experience.

### 6.1.1 Putting Time at Its Place

Before delineating how spatial information produced by mapping performs a function in individuation's process, we need to unfold how mapping works.

Attention should be paid to the term 'map', a very contentious one. Given the impossibility here to adequately resume the debate, the choice has been made of limiting all mentions to viewpoints on 'what-is-a-map' debate exclusively to the authors whose positions are applicable to Simondon's topics. For the same reason here the focus is more on the ways of producing maps than on maps in themselves (albeit their role as *artifacts*, as we shall see, is really relevant, but in Simondon's terms).

First of all, what is the difference between *place* and *space*?

A place is inseparable from meanings peculiar to the living beings dwelling in it. A place cannot be exchanged for another one leaving everything unaltered. Space, on the contrary (particularly, according to the version launched by modernity), is the dominion of metric distance, of measures: in which all places become points which can be displaced because they are equivalent, being all subject to the sole law of the standard measurement system (Farinelli 2009). In this sense, a map is at odds in dealing with places, though being the predominant tool entitled to produce and manage spaces since the beginning of the modern age, and it is currently an influential model. Actually, how could have geographical metaphors for nations come into use—such as the 'boot-shaped' Italy or France's 'hexagon'—without the employment of maps before early Earth images from satellites were made available to the public?

But mapping played a role in the representation of *time* as well.

With the first modern atlas—the *Theatrum Orbis Terrarum* by Abraham Ortelius (1570)—the representation of the present and the representation of the past separate, and it is the present time that gains the primacy: experience becomes more important than authority. And this is clearly shown by two world maps: the *Aevi veteris, typus geographicus*, put in the opening page of the section collecting historical maps, so highlighting the different treatment reserved to the images of the past, all located in a special section devoted to them; and the *Typus Orbis Terrarum*, placed at the opening page of the entire volume.

In the *Aevi veteris, typus geographicus*, Ortelius puts the earth as it was known by the ancients within the global frame outlined by the great planisphere displaying the extent of geographical knowledge in 1570—the *Typus Orbis Terrarum*—but leaving blank all areas outside the ancient known Earth: the simple comparison between the two planispheres was sufficient for the reader to grasp the knowledge gap between ancients and moderns.

Here, on the basis of a geometrical frame *common to both maps*, world's present and past diverge, and time explicitly[2] becomes a social memory, a past to be (or not) adopted.

---

[2] Explicitly, because modernity revealed the customary nature of what was previously considered as a sacred gift.

Summing up: modern age produced a divorce between the representation of space and the representation of time. Such a separation appeared in modern cartographic dress, in which, on the same geometrical basis and in accordance with the tacit principle assigning a determinate temporal frame to each map, maps of past territories and maps of present ones are separately drawn. While medieval *mappaemundi*, according to the plot of Salvation established by the Scriptures, included past, present, and future in a single picture; modern maps, in giving the past a separate space of representation, made the cleavage between past and present visible, perceptible.

In order to more fully understand the power of modern maps it should be noted —as Franco Farinelli (Farinelli 2009) demonstrated long since[3]—that just because modern maps adopted as its basis the space defined by the rules of Euclidean geometry (along with its qualities of homogeneity, isotropy, and continuity), the rich complexity of the world is reduced through them to what can be measured and displayed inside the grid of coordinates, dividing into different spaces of representation the present, the past, the physical features of a land, the political boundaries of a territory: a subdivision that would not have made sense in maps drawn before modern times.

An insightful example, in this regard, comes from European 'Age of Exploration's' maps.

As Arthur Robinson claimed, it was the mapmakers who really discovered America, rather than Columbus:

> It was not until well into the 16th century that the Americas were recognized as continents and thus 'discovered'. The new lands were even named by a cartographer (Robinson 1992: 31).

In a positivistic-oriented vision of the whole history of geographical discoveries, it would be easy to consider any episode of exploration and mapping joined with each other along the causal chain of progress. Indeed, as Robinson put it:

> They [explorers and mapmakers] did so rather like the blind person who tries to make out the form and size of a large object by feeling around it edges. They had to wrestle with the problems of shapes, sizes, and locations of the reported lands and particularly their relation to Asia (Robinson 1992: 34).

But how could explorers find something they didn't know exactly what it was and where it lay? Answer: by the maps themselves. Indeed:

> surely the discovery of something geographical requires that the discoverer have some idea, even if only approximate, of what and where in the world that something is (Robinson 1992: 31).

The explorers were not completely blind, as it were, since they could reduce the complexity of potentially infinite directions to take in order to hit their target. They had a previous image of the world to rely upon. Such an image was provided "by

[3] The first works by Farinelli about the mapping model of Western culture date back to the end of 1970s.

cartographers who painstakingly fitted the bits of geographical information brought back by the explorers into the world map" (Robinson 1992: *ibidem*). Every bit of geographical information reported by the explorers became a trace on the world maps plotted by the cartographers, a trace which, in turn, assumed the role of explorers' memory.

It is easy to recognise here the Platonic aporia (a no-way path) formulated above although in a geographical context: how can we know something, if searching for the known is useless, and searching for the unknown is impossible, since we don't know what we're searching for? Plato's answer is, however, well known: every act of knowledge is but a remembering. But, in what sense did geographical discoverers remember? By means of the world images staged for them by mapmakers. Without such traces no expedition would have ever been conceivable. In this sense, maps (and, at a lower level, literary sources like travellers' books or classical sources) represented the knowledge available at that moment; as, at one and the same time, images to be trusted and to be tested (see here Sect. 6.3).

### *6.1.2 Mapping and Associated Milieus*

Let's go back now to Gregory Bateson's still groundbreaking position.

> All description, explanation, or representation is necessarily in some sense a mapping of derivatives from the phenomena to be described onto some surface or matrix or system of coordinates. In the case of an actual map the receiving matrix is commonly a flat sheet of paper of finite extent, and difficulties occur when that which is to be mapped is too big or, for example, spherical. Other difficulties would be generated if the receiving matrix were the surface of a torus (doughnut) or if it were a discontinuous lineal sequence of points. Every receiving matrix, even a language or a tautological network of propositions, will have its formal characteristics which will in principle be distortive of the phenomena to be mapped onto it. The universe was, perhaps, designed by Procrustes, that sinister character of Greek mythology in whose inn every traveler had to fit the bed on pain of amputation or elongation of the legs (Bateson 1979: 48, note).

Such position makes it clear in what sense mapping is based on *iconic* relations, i.e., as it is commonly presumed, on *similarity*.

Indeed, only assuming that there is a *sympathy*, in the sense of mutual assimilation, between thought and things, between mind and reality, it is possible to relate things according to the criterion of similarity: 'the like is known by the like (…) Plato says that among the similar, friendship is unavoidable' (Melandri 1968: 735).

The map is a model of the world. But models do not work according to the relation of similarity, but of *mimesis* (in the originally Greek sense: Catoni 2008). The differences are basically two: the similarity is a kind of symmetrical relationship while mimesis isn't (things are assimilated into, imitate ideas, but ideas do not imitate things), and mimesis does not only establish the relationship between things and ideas, but it also asserts *grading*, *intensionality* (things approximately imitate ideas) (Melandri 1968: 879 ff). The map would be the product of a *double mapping*:

the first one, which associates the phenomena to be represented with a culturally-defined model of space—which, in its turn, identifies a set of *mappable* elements, i.e., the mapping in the spatial model of the elements being relevant to the code of the model itself—the second one, which associates the set of spatial elements being mapped with the set of elements drawn on the support.

As Gregory Bateson puts it,

> To produce news of difference, i.e., information, there must be two entities (real or imagined) such that the difference between them can be immanent in their mutual relationship; and the whole affair must be such that news of their difference can be represented as a difference inside some information-processing entity, such as a brain or, perhaps, a computer (Bateson 1979: 68).

In current standard cartographic terms (i.e., the Western-generated rules for mapping globalised by colonialism), the content (*C*) (phenomena we want to represent) is translated, formalised, particularly by the rules of Euclidean geometry (*F*), to be translated again into form ($C \Rightarrow F \Rightarrow F'$).[4] Leaving the translation's metaphor, which can be highly misleading, it should be noted that this process, as a whole, is anything but a *transduction*, a crucial Simondon's notion, whose, as for mapping, a good case in point is the mold-clay's example: 'one can say that *the shape of the mold does not work on anything but the shape of the clay*, not on clay's matter (…) the pure form *already contains gestures* and raw material is *ability to become*' (Simondon 1964: 32–33, italics added).

Only what can be homogenised by the Euclidean metric, can be found on this kind of maps. To take shape on the surface of the map, that 'something' is sorted, filtered by the Euclidean matrix. Anything that doesn't pass the filter remains outside of the map. The main Western map's filtering device is the *scale*, but all maps are filtering matrices (not just 'Westerner' maps): it is the kind of filter that change. In this case the filter is the metric scale. Map is compelled to choose the scale first, and such first act decisively determines the information provided. It must be considered, besides, that the Euclidean matrix, in becoming a widely accepted international standard through the metric measure system since nineteenth century, it constitutes mostly the *associated milieu* linking maps to all objects, events, processes, we usually associate with contemporary economic systems (real estate, transportation systems, retail management, and so on).

A corollary is the claim of deriving the qualities, characteristics, behaviours or the identities from the spatial location. If a form can be detected in social actions this happens because of the same nature of social action, based on our beliefs that are adequate if they *fit* the world (they don't *match with* the world) and are woven by rhythmic iterations of deeds and provisional spatial orderings. Conversely, we have no spatial grammar for social action because there is no isomorphism (no *matchness*, as it were) between forms and beliefs, unless we suppose a basic rationality of social action we can compute. An instance of such fallacy can be found in the observations on ethnic homogeneity of Chicago neighbourhoods made

[4] It should be taken into account that any use of temporal expressions is here just metaphorical.

by the seminal sociological school of Robert Park, which have been translated in the zoning of urban development plans as a record of *natural* spatial concentrations. Indeed, these concentrations are regulated by the real estate market.

Of course, the question of what is and what is not a map remains, but, as stated before, here are considered the *ways* of producing maps rather than maps *as artefacts in themselves*. Accordingly, what is really different among cultures in time is not, as rightly pointed out by Denis Wood (Wood 2012), mapmaking in itself, but the historically and socially determined infrastructure chosen.

At this point, it is necessary to stop momentarily for defining what is meant here with the term 'infrastructure', because of its relevance to the Simondonian notion of *associated milieu*.

To Simondon, as it is well known,

> When a new device appears in the evolving series, it will last only if it becomes part of a systematic and plurifunctional convergence. The new device is the state of its own possibility. It is in this way that the geographical world and the world of already existing technical objects are made to interrelate in an organic concretization that is defined in terms of its relational function. Like a vault that is only stable once it has been completed, an object that has a relational function continues in existence and is coherent only when after it has begun to exist and because it exists. It creates its associated environment by itself and it achieves true individualization in itself (…) This associated milieu is the invented technical object's condition of existence (Simondon 1958: 51 and 52).

The current sense in which the term 'infrastructure' is used in the literature concerning the Geo-Web, to which we mostly refer in this essay, it is due, by and large, to the sociology of science, particularly belonging to the edging area interested in knowledge transfer and collaborative works' procedures (Star and Ruhleder 1996; Bowker et al. 2010).

> Common metaphors present infrastructure as a substrate: something upon which something else "runs" or "operates", such as a system of railroad tracks upon which rail cars run. This image presents an infrastructure as something that is built and maintained, and which then sinks into an invisible background. It is something that is just there, ready-to-hand, completely transparent (…) we hold that infrastructure is a fundamentally relational concept. It becomes infrastructure in relation to organized practices. Within a given cultural context, the cook considers the water system a piece of working infrastructure integral to making dinner; for the city planner, it becomes a variable in a complex equation. Thus we ask, *when*—not *what*—is an infrastructure (Star and Ruhleder 1996: 112 and 113).

This also holds true in the case of that particular kind of infrastructure that is the *boundary object*.

> Boundary objects are a sort of arrangement that allow different groups to work together without consensus. However, the forms this may take are not arbitrary. They are essentially organic infrastructures (…) I use the term object in both its computer science and pragmatist senses, as well as in the material sense. An object is something people (or, in computer science, other objects and programs) act toward and with. Its materiality derives from action, not from a sense of prefabricated stuff or "thing"-ness. So, a theory may be a powerful object. Although it is embodied, voiced, printed, danced, and named, it is not exactly like a car that sits on four wheels. A car may be a boundary object but only when it is used between groups in the ways described above (Star 2010: 602 and 603).

In this regard, Simondon has a telling example: the first maps produced in the Greek colony of Ionia in the sixth century before Christ.

Such maps are the solution to the needs of a transcultural society, whose 'operational universality can not be content with translating the cognitive structures of a specific city into those of another one' (Simondon 2006a: 6).

> Above all such private languages that are local cultures, navigators implant a system with the power of the universal, coextensive, as a source of representation, with their ecumenical action, indefinitely extendable. The "Greek miracle" occurred when the mixture of customs, beliefs, religions, techniques made unusable the primitive tools of cognitive representation and communication: those of cities and particular ethnic groups can only represent and translate "adiabatic" transformations, having no exchanges with the outside, with no momentum towards universality; now, Ionian navigators are just precisely the very basis of acceptance of the new; poetry, beliefs, rituals and religions, myths and prohibitions are unable to provide structures of interpretation able to be indefinitely increased and enhanced to welcome and integrate new information brought by travel. A simple cut table where through engravings they are depicted the contours of the shore and the mouths of rivers is worth more, for the navigator, than poetic theogonies; as this table, the symbol of perception, integrates a knowledge that can be accumulated; in the course of subsequent journeys, it can receive new details, inserted among the old ones, and be prolonged beyond the coasts previously explored. On the other hand, *it concretises a universal mode of expression, with no detour through the languages unintelligible to foreigners, more learned and, in a sense, more abstract than poetic or religious discourse, it is also more direct since it is aimed at the individual itself with no detour through cultural learning*. The *pinax* ['table' in Greek] is historically one of the earliest known examples of this transcultural and indefinitely increasable way of thinking, *more objective because less collective than traditional thought*, and thanks to which the cities of Ionia gave birth, in the sixth century BC, to Western philosophy. In this form, and on its debut, philosophical thought is a close relative of perception, since it is the work of lone men, who act without relying on the cultural heritage of cities; their models of intelligibility are operational, constructive, in direct connection with the dexterity of the artisan; the world is conceived, represented, as it might be touched and built (Simondon 2006a: 6–7, italics added).

The more a community needs to trespass the boundary of locale for any reason, the more it shall search for transcultural adjustments of its infrastructures. Extensional and formal infrastructures are thresholds to the *supra-local*.

If we follow such example, it is apparent the role of information infrastructure played by the Ionian maps. Now, leaving in the background Simondon's clever description of maps merging the individual experience and a 'universal mode of expression', it is useful to show *how* this result could be obtained as well. Furthermore, in what sense can they be considered as *milieus* in Simondonian sense?

Taking into consideration the fact that the mimetic asymmetry of mapping is the ground on which Western culture established the relationship between what can be known (things as appear after being mapped *twice*) and what cannot (things-in-themselves) since Anaximander (Farinelli 2003: 30–31 and 164), the character of basic information infrastructure of mapping should be clear. Besides, for information theory, and, in general, for hard sciences (though, recently the difference from soft sciences' approach is getting less and less sharp), information means 'the reduction of uncertainty in a communication system' (Headrick 2000: 3–4), but this implies a human subject interpreting information patterns to make them become human information, i.e., really an available one. Anyway,

> Information is not the same thing as knowledge, though the two concepts overlap. Knowledge refers to ideas and facts that a human mind has internalized and understood: how to fix a flat tire, the name of a really good dentist, speaking French. Acquiring knowledge means absorbing a lot of information—for example, how to use French irregular verbs correctly. Often, the mind acquires and organizes such information in a spontaneous and even subconscious fashion, the way a child learns to speak or a taxi driver knows her way around town. At other times, the acquisition of knowledge requires studying, a slow and difficult process (Headrick 2000: 4).

We can state that, in the following, *intensional* is referred to the production of *knowledge*—requiring a *shared infrastructure* (i.e., not only a common standard of communication, but also a common access to the procedures for changing the standard itself)–, while *extensional* refers to the production of *information*—requiring a high degree of formalisation and abstraction in order to favour its generalisation and diffusion outside the local context.

Comparing the notion of information infrastructure with that one of associated milieu, it is clear their common ground: their primary function being properly to put into relation two otherwise separate *orders of magnitude*. In this perspective, an *invention* is always *localised* (Arthur 2009; Antonelli 2008), as a production of knowledge and not simply of information. Of course, if an invention concerns *supports* (like parchment or paper for writing) it can develop a new level of extensionality, *provided that it become an adopted standard*, so widening its possibilities to put into relation further orders of magnitude, exporting itself outside the original milieu. In this regard, the story of digital support is probably the most renowned.

It has been pointed out that the crucial trait of 'digital revolution' consists in the fact that digital support makes it possible the separation among data, information, and form (Tagliagambe 2008: 79–88), so allowing not only a massive convergence of heterogeneous objects and processes (visual, acoustic, textual, and so on), but also, and above all, the emergence of the *active* role humans play in sensing the world. But this is a genetic feature of maps, which, according to the model of *double mapping* shown above ($C \Rightarrow F \Rightarrow F'$), separately implement data, information, and form.

In this perspective, we can compare Simondon's definition of matter/form relation with what we called double mapping, so refining our model, but again with the caveat that, just as in Simondon's model, the elements compared are not process' stages, it is not a causal chain—with $C$ preceeding $F$ in its turn preceeding $F'$—for this reason, it should be better writing it as: $C \Leftrightarrow F \Leftrightarrow F'$ (Table 6.1).

In the scheme is not the map to be an infrastructure, but mapping in itself. The map is the provisional reification of mapping infrastructure (Star and Ruhleder's '*when*—not *what*—is an infrastructure'). However, in Simondonian terms, if the individuating relation constitutes itself with no temporal succession (there is no "before" and "after"), from the evolutionary viewpoint of the system, time comes into play because each new individuation is based on what was not previously individuated (Neve 2011). Simondon's notion of *pre-individual* is applied here to mapping in the sense of a 'culturally-defined model of space' as stated above, which allows the object-map to be a real information infrastructure, an *invention* (see below Sects. 6.3.1 and 6.3.2).

**Table 6.1** Comparison between Simondon's matter/form relation and *double mapping*

| | | |
|---|---|---|
| C | Matter as potentiality of becoming | States of affairs (objects, processes, relations) |
| F | Pre-individual filtering potentials in transduction + support | Mapping as information infrastructure |
| F' | Form as edging emergence of metastable individuation of matter | Map |

Now, to explain how mapping can be a *milieu of individuation*, it is needed to put the accent on the relevance of Simondon's views on *information* to the process of individuation and the constitution of milieus.

### 6.1.3 Milieus and Individuation

If we reread what Simondon wrote about information (Simondon 2010: part. 59–87, 159–176) in the light of the above line of reasoning, the need of trespassing the locale through maps (as in the case of Ionian navigators) can be meant as a sort of 'potential difference', as a inner system's dissymmetry whose *partial* resolution is its individuation, i.e., its *territorialisation.*

Accordingly, the form is information, but in the sense that is proper to individuation's process. To avoid the fallacy assimilating physical and technical individuation to living's individuation, it should be remembered that, to Simondon, what marks the difference of living's individuation against the other beings is the temporal factor and living's everlasting state of individuation, its metastability, because living's process of individuation concerns, at the same time, the outer and the inner, for the individuation implies also the communication of living being with itself. Then, the living being communicates with itself and its milieu.

This accounts also for the group, the collectivity, as the transindividual individuation of which each member is the amplifying condition. To Simondon there is information if a change occurs as a resolution of the tension between the two dimensions which both individuate themselves by it. In this perspective, information is not a thing, it doesn't concern a unique homogeneous reality, but it is a relation. It concerns two disparate realities in tension, the sense emerging from their individuation as a system. It is the new sense stemming from the new organisation (partially) composing, in a new individuation, the pre-individual incompatibility. If any communication is individuation and that is why forms take place, individuating communication is *informing*, it is information.

Probably, one of the most revealing passages concerning the relation between Simondon's notions of information and milieu is from his well known (though unpublished until recently) conference of Royaumont:

> Local reality, the receiver, is modified in its becoming by the occurring reality, and it's this modification of local reality by the occurring reality that is the function of information. Virtually, it is a receiver any reality which doesn't fully possess in itself the determination of

> its becoming's course (…) The receiver is effectively a receiver if only it is heteronomous from the point of view of the cause triggering transformations (Simondon 2010: 159 and 160).

Milieu it must be here understood in a twofold sense: the milieu as the environment that any living being structures, composes (just like plants or animals), but also as what mediates—in a technical sense, it is a means—so creating a real *in-between world* (Tagliagambe 2008).

Indeed, in geography, the very basic sense of information infrastructure can be traced back to the notion of *milieu* (Neve 1999). The first monograph concerning the philosophy of technique was written by a geographer, Ernst Kapp, a Carl Ritter's disciple (Kapp 1877), and his work was introduced in France by Georges Canguhilhem, one of Simondon's mentors.

> With Galileo and also Descartes it became necessary to choose between two theories of milieu, that is, in the final analysis, theories of space: a centered space, defined as being where the mi-*lieu* is a center; a decentered space, defined as being where the *mi*-lieu is an intermediary field (Canguilhem 1952: 24–25).

The concept of milieu emphasises the dynamic character of places, not static containers of things and beings, but nodes of forces, actions, desires, values. A group of human beings, to live in a certain environment, *processes* its own world, in which territory is the techno-symbolical medium, being it, at the same time, the technical response to the needs of the group with respect to environmental factors, and the materialisation of the idea allowing to elaborate that response, and to develop it or reject it later. Every human generation is born within a given technological horizon which constitutes his technical milieu of reference—the way of the water to fish: living environment yet not perceived (Aristotle, *De Anima*, B 11 423 a 31–423 b 1).

In this sense, it should be evident why a map is not simply a map, just as objects are not simply objects.

> It is an essential characteristic of being human that we feel the urge to reify experience, to give those fleeting moments of pleasure and pain a narrative outline or visual shape (…) Artifacts are not normally objects of thought. As specialized tools artifacts extend our sense of competence in the world; we do not think about them as long as they continue to function. As multipurpose dwellings they stabilize experience, support our sense of a perdurable self, and confirm our belief in a comprehensible universe (Tuan 1980: 462 and 471).

There is a passage from the *Odyssey*, which portrays the prophecy of the soothsayer Tiresias which Odysseus listens during his pilgrimage to Hades, and then which he repeats to Penelope (*Odyssey*, XI, 119–137): when Tiresias says Odysseus, after having killed the Suitors, he shall

> pick up a shapely oar and travel on till you come to a race that knows nothing of the sea, that eat no salt with their food, and have never heard of crimson-painted ships, or the well-shaped oars that serve as wings. And let this be your sign, you cannot miss it: that meeting another traveller he will say you carry a winnowing-fan on your broad shoulder. There you must plant your shapely oar in the ground, and make rich sacrifice to Lord Poseidon.

One may designate the same object belonging to a culture of the sea, like oar, or to a culture of the land, such as shovel to separate the chaff from the grain, so marking the border where techniques and technical marine objects are taken for techniques and technical objects of the mainland. Of course, the shape of the oar is not helpful. Indeed, it is precisely the fact that it presents no unique information, that his form is not readable without ambiguity, to make it possible that the oar is mistaken for another object. The shape can be tricky. Objects that crowded the world before the industrial revolution (and still living with their versatile successors) lent themselves to more than a single use from the same form: after all, an oar could be mistaken for a winnowing-fan because it could also be *used* as a winnowing fan. The bare presence of a new technical object defines a spatiality which becomes an orienting milieu for subsequent generations. The object is memory, retaining in its praxis a knowledge that is part of the history of the places that it has individuated. You can not separate the winnowing-fan from wheat's fields, from the seasonal rhythms of cultivation and climate, from ownership contracts and the trade of products, from warehouses, from the inns, from harvest festivals, from the prayers of thanksgiving and imprecations, dialects and dishes.

If a technical object exists only if it creates its own milieu, it follows from all said above that for a map the "stuff" of its milieu is information. And the recent developments of Geo-Web mapping underlines this trait even more. So it should not be surprising to read the following definition of mapping made by two of the earliest scholars who attempted to investigate the new issues Web was raising since its debut (Dodge and Kitchin 2001).

> For us, maps (…) have no ontological security, they are ontogenetic in nature. Maps are of-the-moment, brought into being through practices (embodied, social, technical), always remade every time they are engaged with; mapping is a process of constant reterritorialization. As such, maps are transitory and fleeting, being contingent, relational and context-dependent. Maps are practices—they are always mappings; spatial practices enacted to solve relational problems (…) A map is never a map with ontological security assumed; it is brought into the world and made to do work through practices such as recognizing, interpreting, translating, communicating, and so on. It does not re-present the world or make the world (by shaping how we think about the world), it is a co-constitutive production between inscription, individual and world; a production that is constantly in motion, always seeking to appear ontologically secure (…) From this perspective, the important question is not is not what a map is (a spatial representation or performance), nor what a map does (communicates spatial information), but how the map emerges through contingent, relational, context-embedded practices to solve relation problems (their ability to make a difference to the world); to move from essentialist and constructivist cartography to what we term emergent cartography (Kitchin and Dodge 2007: 12).

## 6.2 Evolutionary Trends of Milieus

Anyway, to account for the real novelty represented by the so-called Geo-Web and its revealing effect on the function of mapping as a milieu, we must mind about two elements belonging to the evolutionary framework of human milieus.

The former concerns the separation between *knowledge* and *practices* as a character of modernity coming to an end nowadays (at least in information-driven activities); the latter constitutes a basic anthropological trait relevant to individuation's issues.

### *6.2.1 Knowledge/Practices*

To avoid confusion, it should be clarified, preliminarily, an apparently internal contradiction. The definition of knowledge previously given (Sect. 6.1.2), as opposed to *information*, would seem to not match the following one, as opposed to *practices*.

Taking into consideration the fact that such oppositions are to be regarded as not being absolute (knowledge is also "made of" information), the difference between the two definitions does not affect our argument, since what is at stake here is not the *complexity* of knowledge in itself—depending on 'being in a world that is inseparable from our bodies, our language, and our social history—in short, from our *embodiment*' (Varela et al. 1993: 149)—but an historical threshold, which is responsible for having implemented the dichotomy between *explicit* and *tacit* knowledge (Polanyi 1958; Wenger 1999), while enhancing the former ('knowledge *that*') to the detriment of the latter ('knowledge *how*') (Varela et al. 1993: 148).

> According to tradition, the "practice" is actualised, while "knowledge" ensuring its consistency is potentalised (…) We lived in the past everyday life on the traditions of the body, of nature, and society. Traditions that fueled action and framed work (…) Yet such tradition mobilised by work was closely linked to conditions determined by the eco-logic and socio-logic of a place, a time, and a group. Changes occurring in these "logics" could make tradition ineffective, and practices hitherto successful came into crisis (…) In other words, tradition made it difficult to cope with unexpected events, the territoriality torn asunder in its depths could only result in a tragic everyday life… until things go or find a new balance. The confusion of knowledge and practices prohibited to imagine new practices from explicit knowledge (…) The explicit knowledge or if you prefer the dichotomy between knowledge and practices is precisely what characterizes modernity which can be defined as a process of fission characteristic of Western thought (Raffestin and Bresso 1982: 188).

The massive and still growing development of knowledge since the Scientific Revolution mainly owes to such dichotomy, along with its awesome technological supplement—"technological" above all in the sense of the subsequent (and still in good health) hookup of technology research with industrial production. In fact, such development should have to favour mutually compatible standards and information infrastructures (Simondon 2008: 177) in order to allow different inventions to take place in constituting 'techno-geographical milieus', i.e., *territorialities* (see the example of railroads, in which mapping and time coordination strictly cooperated with each other: Galison 2004).

As a matter of fact, tacit, local knowledge was not eliminated, as it is the very ground on which knowledge is generated, even in the case of scientific inquiry:

> Scientific knowledge gleaned in laboratories is thus less about the local instantiation of universally valid facts than about what one writer calls 'the adaptation of one local knowledge to create another' (…) To put it another way, the reason a person gives for behaving in a certain way is setting dependent. This means that standards of practical rationality—what passes as a good reason for believing something—are spatially referenced (…) Rationality is always *situated* rationality. And it is always embodied rationality (Livingstone 2003: 142 and 184).

The *predictability* is the great power gained by humans in enhancing explicit, formal knowledge. A power by which they could afford to overcome many of the uncertainties which local lore's 'logics' were unable to face. Such predictive character of explicit knowledge is granted by the use of models, 'powerful instrument of an explicit and formulated information being, for thinking, like the tool that everyone carries with itself and uses by its own forces' (Simondon 2005: 78), just like mapping.

On the other hand, as already said, the production of knowledge remains *localised*, and could not be otherwise.

> The differences between the building of Chartres cathedral, for example, and modern technoscientific practice lie not in the possession of some secret or mysterious skill nor in some essential difference between science and technology on the one hand or between theory and practice on the other. Both science and technology, now and in the past, are the product of local and tacit knowledge. The differences between them lie in the social and technical means by which local and messy knowledge and practices are made robust, coherent, and mobile, that is, in the ways in which the site-specific or even problem-specific products are added to the work of previous individuals or groups of workers or transmitted to another site (Turnbull 1993: 317).

It is not by chance that recent insightful views on teamworks' organisation (not only for research) draw upon the dialectics between *reification* and *participation* (given that, as for knowledge and practices, they are not mutually exclusive):

> Explicit knowledge is (…) not freed from the tacit. Formal processes are not freed from the informal. In fact, in terms of meaningfulness, the opposite is more likely (…) In general, viewed as reification, a more abstract formulation will require more intense and specific participation to remain meaningful, not less (Wenger 1999: 67)

It also holds true that Web's overwhelming emergence as a milieu of *collaborative, distributed* knowledge (Benkler 2006) is outlining a probable new *convergence* between tacit and explicit, practices and knowledge. And on this trait new forms of mapping are grounded (Sect. 6.3.2).

### *6.2.2 The Aspect's Doubleness and the Pre-Individual*

Simondon calls the surfeit non-individuated, as it is known, *pre-individual*. This is the pre-individual reality prior to any individuation, an individuation whose each stage is never fully accomplished, being in metastable equilibrium. The pre-individual presents itself both as 'a kind of unresolved past' and as 'the environment of the individual identified. The environmental context (perceptual or linguistic, or historical), which is part of the experience of the individual, is, in effect, an intrinsic component (if you prefer: inner) of the subject. The subject does not *have* an environment, but *it is*, in a certain part of itself (the one not individuated) environment' (Virno 2001: 236).

In such perspective, how does it work the adaptation of humans to the environment, their fitness? It is based on the mismatch between perception and action.

> We exist as we do not coincide with ourselves [...] we are local and located by biological and cultural necessity and global by socio-historical evolution (Morelli 2011: 80 and 55).

From an evolutionary point of view, the human animal, since it does not live in an ecological niche, is constantly pressed by a surplus of informational stimuli always exceeding the amount of information strictly necessary for self preservation, the reason why human milieu is characterised by uncertainty and potentiality of meaning. One must interpret and, therefore, choose a path, because 'the glut of impressions and solicitations doesn't result in a detailed catalog of vital requirements'. All this implies a necessary distance of the human animal to the environment informing the *doubleness of its aspect*: that is to be 'a naturally artificial animal, namely an organism whose biologically distinctive trait it is culture'. Such *distancing* also seems to have been proved on a neurological basis (Libet 2005).

The resulting representation of reality—being not directly correlated with perception—is validated by a *meta-representation*, 'which is the substitute of the stimulus, by assuming the orienting function that it does for other species' (Virno 2010: 30, 31, and 33).

However, the fact that the human animal – due to its own natural artificiality—lives in an excess of information cannot be solved in a simplistic opposition between the human world and the environment. In reality, the human animal is constantly trying to reduce the informational indeterminacy by its historical-social acts, producing 'historical social niches' (Virno 2010: 39). This situation can be defined, in Simondon's terms, as *metastability*.

In this way, it contrives habits and practices, codes, in other words, it *actualises the world* (reducing the potentiality of meaning), so ensuring a proper ratio between information and actions. But this ratio, in turn, it is not stable, because it is the product of historical and social choices, and a crisis of information's reliability is still possible. This crisis manifests itself as a *virtualisation* of environment, in a word, a *deterritorialisation*, often putting back on track paths denied by previous choices (increasing potentiality again).

In this sense, the focus is on dynamic relations and the role of the technical milieu in translating (not simply conveying) feelings, memories, expectations, into actions and beliefs:

> Translating is not an idle occupation for a limited circle of specialists, it is the texture of everyday life, the work that we perform each day and each hour of the day. We are all translators, since translation is the property common to all forms of life. It is a necessary aspect of being-in-the world in the "information society". Translation is there in every form of communication, in every dialogue. It must be so, because the plurality of voices cannot

> become eliminated from our existence, which amounts to saying that the borderlines which dissociate and fix meanings will continue to be drawn in a fragmentary and spontaneous manner, even without any superior office of cartography and any official maps from the Geographical Institute (Zygmunt Bauman, cit. by Balibar 2009: 205).

Indeed, we can think territory in relational terms, as a set of relations concerning times and places, relations which, often, cannot be mapped properly.

Thinking in terms of territoriality enables us to reconsider contemporary fragmentation of places' experiences, rethinking conflicts in a broader framework.

> To "territorialize" means to assign "identities" for collective subjects within structures of power, therefore to categorialize and individualize human beings (and the figure of the "citizen", with its statutory conditions of birth and place, its different sub-categories, spheres of activity, processes of formation, is exactly a way of categorizing individuals). Such a process is possible only if other figures of the "subject" are violently or peacefully removed, coercively or voluntarily destroyed. It is also always haunted, as it were, by the possibility that outsiders or "nomadic subjects", in the broad sense, resist territorialization, remain located outside the normative "political space", in the land of (political) nowhere which can also become a counter-political or an anti-political space (for which Michel Foucault coined the expression heterotopia) (Balibar 2009: 192).

The archaic Italian expression "comune sentire", which was once a synonym for public opinion, can therefore be understood, in this sense, as "common sensing", as a perceptual sphere *in common*. Urban space is *synesthetic* in sofar builds *a space of common sensitivity, a space of feeling together*. And it is the materiality, the reified ideas, *cultural traits* (Stone et al. 2006), to be the condition of possibility of such sharing. The cultural characteristics, such as "social objects" (Ferraris 2009), materialize the relationships among individuals and constitute a social memory (in this sense reification is very different from *fetishism*, which is a relationship *between* things). As summarized by Hannah Arendt:

> To live together in the world means essentially that *a world of things is between those who have it in common*, as a table is located between those who sit around it; the world, like every *in-between*, relates and separates men at the same time (Arendt 1958: 39).

## 6.3 Networks, Mapping, Milieus

Today, we experience new relations between places and times, experiences mostly driven by the development of media, particularly ICT.

> Different time horizons and different rhythms or tempos have always existed in society, particularly in modern societies. These are differentiated into functional sub-systems whose boundaries allow for the development of different temporalities and tempos in their respective systems (and their organizations). They are synchronized and integrated by institutional arrangements and—in the Parsonian version of systems theory—by the cultural system, that serves the function of latent pattern maintenance (...) The (different) speed(s) in the different systems thus could be controlled by the 'land factor' of culture. In the diachronic dimension, the different times can be turned into the continuity of tradition, and

> in the synchronic dimension the normative structure is ordering the co-existence and succession of commitments and events (…) But in our contemporary society, something seems to be different. The synchronization of the very different tempos and temporal horizons seems to function well, if they are existing apart together, within the boundaries of their respective subsystems and organizations, but integrated by an institutional regime and a temporal ordering of events (Brose 2004: 16–7).

It is apparent the role of culture, of education, in helping individuals with their efforts to blend pasts and places. And it is no accident that maps played a prominent role in national education systems.

In our societies the space of media (particularly, taking into consideration the global convergence to digital) intensifies, at a continually increasing rate, the complexity of places' relations.

But it is the declining role of the 'land factor' of culture which is our main concern here. The fact that the bond between places and cultures (see Sect. 6.1) has been put in crisis at first by the extension of transportation systems during 19th century, and almost cut later by the electronic media of 20th and 21st centuries, foreshadows a situation of cultures, i.e., pre-individual, circulating globally with no links with their generating contexts.

It is the phenomenon Elmar Holenstein describes as the 'giving out of the mediation in intercontinental connections' and 'de-regionalisation of cultural differences' (Holenstein 2004).

This would mean that it is possible to gain access to virtually any culture from any place without moving, but we know, from Simondon, that pre-individual is never fully individuated, and the resulting metastability fuels the search for an integration into a higher collective level, a resolution of tension through the transindividual, i.e., in this case, the 'contemporary informational sets' (de Boever 2012: 231) of which the Web is the most renowned (and probably the most influential).

> His [Simondon] key idea is that any individuation remains dependent, in a metastable equilibrium, on the *pre-individual potential* from which the individual emerged through successive 'structurings' or 'distanciations from the environment'. Therefore the existence of an individual is always "problematic" or tense. It is this tension which individuals try to resolve (or to understand) by seeking a greater degree of individualization in the building of collectivities. But a living collectivity is never a simple aggregate or, on the contrary, a fusion of pre-existing individuals: it must be a *culture* (what Simondon also calls 'spiritualité'), or a dynamic way of solving the individual's problem. It has to return to the pre-individual level (consisting, among others things, in emotional patterns) to integrate them in a new, superior, metastable entity which, for that reason, will appear as *neither "external" nor "internal"* to the individuals (but precisely transindividual) (Balibar 1993: n. 24).

"Culture-from-anywhere" does not mean, therefore, "culture-as-a-whole", since, in order to enter transindividual dimension, it is to become another thing: using the terminology introduced above, it should gain in *extensionality* while losing *intensionality*.

As a matter of fact, as noted by Olivier Roy with regard to religions:

> If religions are able to extend beyond their original cultures, it is because they have been able to "deculturate" themselves. The religious marker circulates without cultural markers, even if it means reconnecting with floating cultural markers—*halal* fast food, eco-*kosher*, cyber-*fatwa*, *halal* dating, Christian rock, transcendental meditation (...) The deculturation of religion has some fundamental consequences: first of all it transforms the gap between the believer and the non-believer into a barrier, since now they no longer share either religious practice or common values (...) Deculturation is the loss of the social expression of religion (...) On the other hand, the simultaneous presence in the market of different "religious products" results in both competition and standardization, not of theology but of religiosity (Roy 2011: 7–8).

### *6.3.1 New Places, Old Situations*

At this point, we can examine anew the question of place-sensing in the light of the lines of argumentation developed till now. We will draw on the notion of "invention" as elaborated by Simondon to analyse, at first, in what sense place-sensing is encountering a change, and, finally, to unfold the issues raised by Geo-Web.

It has to be remembered that the convergence of telephone and computer is recent history and with plenty of innovative leaps (the first satellite communications equipment is operating in 1960, the first personal computer in 1981, the first digital telephone networks come into use in the early eighties, the Internet—previously dedicated to military security functions—began to work as a system for documents' exchanging in 1989, and in 1993 the World Wide Web becomes publicly operating and accessible). The ability of the Internet to create this amazing blend of interconnections between different systems and media (telephone networks, routers, IP addresses, backbone etc.) giving any home terminal (and now also mobile: smartphones and tablets) access to virtually all the information available on the networks, it is especially explicit in the adoption of common protocols for the transfer of "packets" of information and in the conversion toward digitisation, through which texts, images, sounds, can all be merged in a single information palimpsest available to the user.

Networks' working, as studied in information science while showing a more and more intense resonance with other fields of research (Easley and Kleinberg 2010; Barabási 2009; Bowker et al. 2010; Damasio 2003; Hendler et al. 2008; Miconi 2011; Shadbolt and Berners-Lee 2008; Tagliagambe 2008, 2011), is gradually foregrounding (and, in a sense, actualising) issues concerning what Simondon meant by 'transindividuality', specifically by letting new milieus emerge, leaving the '"modern" human society of work—which was born from the industrial revolution, and which was made up of merely *inter*individual relations' to enter a dimension of '*transindividuality* that is indissociably human and technical' (de Boever 2012: 231). This poses a series of questions about the "behaviour" of networks. For example: to what extent do networks "mirror" users' behaviour, so constituting a reliable model of human conduct? Or: can the sense of "place" that many aspects of networks are implementing still be called "a feeling for places" (above all in the case of the interplay linking the fields of Human-Computer Interaction (HCI), Computer Supported Cooperative Work (CSCW), and Interaction Design: Ciolfi 2011)?

To Simondon, invention is a 'communication between two orders of magnitude, that of the result (…) and that one of the Event-problem' (…) it is the selective recruitment of certain data from past experience made by the current representation of the concrete end to be reached. The invention, as organisation, it is here a detour through the past' (Simondon 2008: 140–141 and 151). Inventing is to identify a relationship. In spatial terms, it means to locate imaginatively the "analogical space" that the current situation we encounter could share with the situation that restores the continuity of action, releasing us from the discontinuity caused by the problem. To ensure that the solution not be just contingent, local, it is necessary that the solution be incorporated into an "object" enduring contingency. This makes it possible to generalise such solution and makes it applicable to different times and places.

An invention like the Web does selectively take data and information from the past implementing them as gateways to future situations. The example of *skeuomorphism* should suffice (above all for its relevance to Web's would-be capacity to generate places).

The technical term "skeuomorphism" refers to a kind of objects' design imitating another material or technique. While such term dates from the end of nineteenth century, it is now commonly employed as to computer interfaces. In fact, users (even more since touchscreens' massive spreading) "are in touch" with *analogical* interfaces simulating common objects and experiences (desktop arrangements and tools, scrolling pages, and so on), while the real working of machine is ruled by *digital* languages. This is an old solution, indeed, because all new industrial objects are "introduced into society" by a "familiar" interface to be user-friendly (first typewriters granted by patents in 1868 had a keyboard which looked like a piano keyboard, as piano was a very common object to be found in middle-class houses, so making the approach to the new invention easier). But it is on this ground that web sites can be experienced as "places", especially in cases like those of so-called virtual worlds like Second Life. It should be noted that once a technical object becomes accustomed, as a part of everyday routine, and widely shared publicly, it occasionally leaves skeuomorph design solutions to implement new features being hindered by the previous design (think about the iPhone, which in the next planned version should give up all skeuomorph interface's features).

But, as to the issue of places "generated" by the Web, 'places-by-proxy' as one might called them (Casey 2002: 51), there is more. In this regard, we will briefly mention the case of 'augmented reality'.

With 'augmented reality' is meant a new way of designing technical solutions employing Interaction Design's techniques to 'potentiate' the real-time experience of a real-world environment by computer-generated sensory inputs. It is used, e.g., for entertainment (video games), but it is acquiring great prominence in cooperative work and scientific research: for example to build a working model of a healthy human organ to compare it in real time with its ill analogue belonging to a patient.

This is very different from traditional models, as, in the case at hand, this model is not a reduced representation of reality, nor something "added" or "superimposed" to reality, but something which *integrates* reality, becoming part of the context, by enhancing some aspects reality already has, though potentially (the hearth's model works like the patient's heart should work if it were healthy, and it allows to operate the ill heart with an improved awareness of body's general situation). In a sense, it *actualises* some potential aspects of reality, so representing a good instance of what Simondon defined as the *inner* and *outer compatibility* of an invention.

In so doing, it permits to 'act in the real world, fully enhancing certain objects' useful characteristics, selecting the most useful ones than those less adequate as to the aims to be reached, potentiating the "collaborative response" environment can provide toward specific needs of subjects being immersed and acting in it" (Tagliagambe 2008: 87).

Augmented reality is not a virtual one, like Second Life which is a totally fictional world, this is *part* of the real world. So, what kind of "places" are they? In a sense, places like these are quite "natural", if we recall what said above as to the 'natural artificiality' of humans. Actually, the issues "augmented places" raise would rather concern the evolutionary trends we mentioned above (Sects. 6.2.1 and 6.2.2).

When we sense by our body, it is an *educated* body which is at work, and *enaction* (i.e., the cognitive action implying that 'perception consists in perceptually guided action and cognitive structures emerge from the recurrent sensorimotor patterns that enable action to be perceptually guided': Varela et al. 1993: 173) linking us to milieus (indeed, *generating* our milieu), it cannot avoid our pre-individual: the stuff that we (mostly) are made of. Augmented reality takes place from certain traits or features reality already has, though often potentially, and in so doing it actually works on some aspects of pre-individual (anything that can be used to describe an heart's model 'in good health' belongs to the given relative knowledge available at a given time). The reduction of uncertainty it operates—*actualising the world* (2.2)—does not work like mapping but by implementing a 'collaborative response' by the milieu. This means at least two things: that the way pre-individual is usually accessed is here affected by a radical change (unavoidable but hard to be envisaged); that the role of the *preliminary selection* of the elements to be "augmented" (what in mapping is meant to be coded instead) gains an even greater emphasis.

The issue at stake here then does not concern a fictitious opposition between a "natural" and an "artificial", or "cultural", approach to the world (as we are 'naturally artificial' animals), but rather it is an affair of purposes, of choices. Places are geographically made by subjects, and, if we examine the original meaning of the Greek word *ethos*, i.e., the home, the place of stay, the dwelling, we should not be surprised to find that places are *ethical*: 'ethics is a way to make the world habitable' (Besse 2000: 144).

### 6.3.2 New Mappings?

In bringing to an end the present essay, we will re-examine mapping's issues in their recent developments driven by the growing trend of 'map mashup' and by the confidence in GIS based on 'ontologies' revised in the light of Simondon's lessons; finally closing by reconsidering the question of *placeness*.

The main issues to be discussed in the last section of the essay, though to a limited extent, are: what kind of maps mashups and GIS are? (of course, this implies the more fundamental and contentious question about what a map is, a question which can not be dealt with at length here); moreover, how the theme of spatial information we left previously undefined (Sects. 6.1, 6.1.1 and 6.1.2) is it managed and processed by would-be new mapping?

As stated above (Sect. 6.1.2), map is based on mimesis, which implies asymmetry in the relation between what is to be represented and the representation; besides, such relation ranges, varying in accord with the codes of the culturally-defined space on which representation is structured.

The two dimensions that any map must harmonise are the local, situational, contextual, indexical, and the transcultural, transindividual. In fact, the transcultural, the formal, the abstraction are the keys to generalisation, i.e., the *deterritorialisation*. So, local and transcultural information, as they are mustered and merged by the map, reflect the condition of metastability between enhancement/reduction of complexity (Sect. 6.2.2). It is this the conceptual node around which revolves the seemingly plain nature of maps (Wood and Fels 2008), as a run-of-the-mill tool everybody could use and believe to know. Indeed, as we tried to show with the help of Simondon's insights, the issue is a real complex one, grounding on the basic and often misinterpreted theme of *analogy*:

> The semantics of the models, which would seem to be a pure question of semiotics (the point *P* of the world map should be read as a sign of the place *P′* of the world, and so on) requires a symbolic interpretation of the model: we must first understand the projective metaphor for which the world map stands for the map of the world (Melandri 1968: 878–879).

In 1979, David Russell, in a groundbreaking yet little read essay, claimed that 'if we regard geography as a theoretical language, we can perhaps usefully think of any map as a particular model sentence in that language' (Russell 1979: 339). In 2012, Denis Wood, following other ways, comes to define the map on a ground being common to what we previously defined (Sect. 6.1.2) by the relation $C \Leftrightarrow F \Leftrightarrow F'$ (take into account that this is a *transduction*: i.e., information—in the sense of "in-form", taking shape—not stemming from the form/matter's relation, but from the spread of a structure on its own border within a pre-individual domain).

> The map had first to declare, insist upon, vouch for, postulate, or propose that the things *were*, that they in some way existed. The map had to say of each: *this is*. Fels and I think about these declarations as *precedent existential propositions*, where a proposition is simply an affirmation that something is or is not (Wood 2012: 289, italics in the original).

Moreover, Wood claims that, as maps, in referring to existing locations, instantiate conceptual types (for example, 'bay', 'river', 'town', and so on)

> the 'thereness' acquires a 'thisness' at the same instant that the 'thisness' acquires a 'thereness'. The simultaneous assertion that 'this is there' and 'there is this' constitutes the posting [i.e., the instantiation on the map] and postings are the 'what' out of which maps are built (Wood 2012: 292).

This is what Wood calls the 'fundamental cartographic proposition',

> Whatever the *this-is* is exists *there,* for example, this border (between France and Germany) exists there (where we draw it on the map); just as it insists that the border is *there* where the map says it is (Wood 2012: 292 and 296).

But this can work like that because the 'posting' puts into relation conceptual types with toponyms, i.e., proper names, as the very ground on which the *extensional* efficacy of the mapping is built. As Franco Farinelli pithily sums up by means of Wittgenstein's *Tractatus*,

> 'In the proposition the name represents the object'. (3.22). That is 'The name means the object. The object is its meaning.' (3.203). Which is the first and fundamental rule of cartographic logic (…) Only on a map, one thing there is or not, there exists or does not exist: unlike what happens in the language (…) *tertium non datur*. This is the second rule of cartographic logic (Farinelli 1992: 9 and 11).

That mapping be a twofold device, merging two sides—logical and analogical—allowing it to mediate between *intensionality* (Wood's *thereness*) and *extensionality* (*thisness*), it is an issue we can not tackle here at length. Suffice it to say that it is the "logical" side which has gained primacy in recent mapping developments; even though, before the arrival of 'map mashup', a great contribution to the maturing of new mapping has come from *network visualization* (Lima 2011), a promising research domain which, unlike *graph drawing* and the attempts to map cyberspace (Dodge and Kitchin 2001), due to its real global viewpoint, it can not avoid to face both "sides" of mapping.

When in 2005 Google Maps went online, they were hacked, mixing (the term 'mashup' comes from the musical jargon) different websites' functionalities or developing open source softwares, so producing 'map mashups'. Employing a specific programming language (XML), which can encode documents being readable by humans and machines, and public application programming interfaces (API), map mashups can generate web-based customised interactive maps. The proof of the pudding of the new public web tool Google cleverly let netizens to develop occurred shortly after the release of Google Earth (GE), the most renowned and the second most used interactive digital imagery of the entire Earth. In fact, when Hurricane Katrina hit New Orleans, it was on Google Earth that millions of people could find the information not available on media. Even US government agency NOAA preferred to post its imagery along with that one from user community, so obtaining in almost real time the evolution of the situation.

Now, map mashup, above all after Google's releasing of 'mapplets'—i.e., utility programs allowing to annotate Google Maps and to add them various applications—, it has become a kind of "map-it-yourself" tool publicly accessible, and so user-friendly to be considered a 'little GIS' (Crampton 2010: 28).

As it is well known, GIS is an information system, whose digital data are geo-referenced, so permitting to store, analyse, and query interactively a database location-related, as well as to display the results in maps. GIS, as the state cartography of nineteenth century, until map mashup's arrival, it was an highly specialised field of research and development, reserved for governments and large corporations, both because of the financial and technological resources needed and the complexity of the softwares used. As a matter of fact, the first GIS dates back to the sixties and it was implemented by the government in Canada for inventorying land use (Crampton 2010: 69).

Map mashup goes in the opposite direction instead. It appears as the first huge new convergence between knowledge and practices (Sect. 6.2.1) through mapping. What it seems to promise is the chance for everybody to produce and share customised spatial information, and not simply for trivial uses like those are usually found on social networks (Cherubini 2008).

But GIS, as developed by great research centres, is going ahead, in employing a new web tool, the Semantic Web, which, simplifying, through a new method for modelling information called Resource Description Framework (RDF), can give meaning to the glut of Web data, assigning a uniform resource identifier (URI)—analogue of the uniform resource locators (URL) identifying web pages—to web resources (i.e., anything identifiable on the Web). Such complex programming frontier has been defined as a gateway to the foundation of a 'Web Science' (Shadbolt and Berners-Lee 2008; Hendler et al. 2008).

> The semantic Web will allow programmers and users alike to refer to real-world objects—people, chemicals, agreements, stars, whatever—without concern for the underlying documents in which these things, abstract and concrete, are described. While basic semantic Web technologies have been defined and are being deployed more widely, little work has sought to explain the effect of these new capabilities on the connections within the Web of people who use them (Hendler et al. 2008: 68).

Now, RDF relies on 'formal ontologies' (Kavouras and Kokla 2008), concepts' descriptions grounded on so-called 'triples', resembling to the "subject-predicate-object" of natural language; so trying, in programming language, to reduce the uncertainty of queries progressively instantiating conceptual description just as map's *posting* does with regard to mapping conceptual types. But, again, we have a mismatch between 'knowledge *that*' and 'knowledge *how*' (Sect. 6.2.1).

Indeed, if it is true that '"ontologies" stabilize knowledge through abstract triples (semantic definitions) to make GIS interoperable', it is also true that, as acutely noted by Jeremy Crampton,

> a GIScientist can assert that a flood is a weather phenomenon but cannot answer the question of what it means to be a flood. This raises serious questions about the ability of "ontologies" (ontical knowledge) to say anything meaningful about lived experience and being. On this account, we don't learn anything about being in the world by abstractly staring at something and listing its properties. Rather, we need to *encounter* the world in its being (Crampton 2010: 108).

With a keen sense of irony, Crampton shows the limits of such method spotlighting ontologies' features in comparing them to the behaviour of a patient of renowned neurologist Oliver Sacks, affected by a disorder which prevents him to understanding what an everyday object in his hands *is*, getting lost in endless description of its *properties*:

> 'A continuous surface' he announced at last, 'infolded upon itself. It appears to have'—he hesitated—'five outpouchings, if this is the word'...later, by accident, he got it on, and exclaimed, 'My God, it's a glove!' (Cit. in Crampton, 2010: 109).

As to *placeness*'s issue, ontologies' method seems like the recent attempt to find 'what makes Paris look like Paris', that is, to use the resources of geo-coded information to make Benjamin's *aura* (Benjamin 1968) computable through 'a large repository of geotagged imagery', seeking 'to automatically find visual elements, e.g., windows, balconies, and street signs, that are most distinctive for a certain geo-spatial area' (Doersch et al. 2012).

And it is from this clue that we can reach the conclusion of the present essay, bringing together all argumentative strands to a possible direction of thinking that could avoid the fallacies listed here.

Michael Benedikt, who was one of the first scholars to really grasp the importance of the Web (though not being an optimistic pundit like the many who flock to this day the media) was searching, since the end of the seventies, for a method to make places' spatial information computable, identifying *space* with *information*.

> The idea of information, of "data", already contains the idea of space within it and vice versa: that there seems to be space in this world at all, even "physical space", is because the world produces information, indeed *is* information (Benedikt 1996).

Benedikt was well aware of what we called here, following Virno, the *doubleness* of human animal before the information overload. In fact, when he provides an example of placeness and the reason why humans need to distantiate them from the environment, he portrays a small place: an open air market in Rome:

> I am a dam, a delay; and this information is accumulated behind me and in me like a lake. A world is constituted in me because the world cannot flow though me un-judged, unreduced. I can respond with actions and thoughts to a vanishing fraction of the welter of information I can perceive, and I "perceive" only because I cannot respond fast enough. Think: were it not this way, were to I have a specific and instant response for every bit of data entering my system, then I would be nothing but a "throughput" device with a passive display (...) In other words, just as stomachs make it possible for creatures not to eat all the time, so consciousness makes it possible for us not to react all the time, to hold a world behind my eyes and posit it before my eyes...while I decide what do next (Benedikt 1996).

To Benedikt, that is why we have a 'spatiotemporal consciousness of a "world"', i.e., the 'frames' we examined above (Sect. 6.1), but he thinks this is also the basis of Lefebvre's *social space*, as a net of information interchange.

Now, on this ground Benedikt develops a theory of *isovists*, meaning by such term the 'set of all points in an environment of opaque surfaces that are visible to a given point *x*'. Provided that environment be built, isovists provide at least two undoubted gains over the usual coordinates' grid: they incorporate motion's factor, then the ability to obtain *time-related* spatial information's measures; they define the *uniqueness* of locations ('every point *x* in an environment generally has, belonging to it, a uniquely-shaped isovist'). But they do so not referring to an *allocentric* spatial grid (Sect. 6.1), but rather to what is *visible* from each point *x*, so drawing closer to the living experience of places.

It could be objected that all place-sensing can not be reduced to the visual dimension (Benedikt himself is not unaware of that). But we want to focus on another aspect of such stimulating viewpoint: the attempt to make spatial information computable by the mutual "mirroring" of extrinsic and intrinsic dimensions. i.e., by a measurement merging *allocentric* and *egocentric* frames.

> Isovists provide us with a way of understanding space itself as neither logically prior nor subsequent to isovists and the information they contain. Isovist measures constitute a class of information about (visual) space. But it is equally true to say that (visual) *space is partially constituted by the structure of, and in, isovistic information*. Extrinsic and intrinsic dimensions turn out, in this case, to mirror one another. We can neither trust one more than the other, nor claim one to be more real than the other. Isovist information is embedded in/at every point in space, *as* "that point in space", *as* space (Benedikt 1996).

Now, while it is a desirable goal to pursue the search for a method to measure spatial information avoiding the fallacy of an allocentric-centred approach, one should be worried about that sort of double bind always lurking in the shadows of such attempts.

> The significance of a basic medium to its civilization is difficult to appraise since the means of appraisal are influenced by the media, and indeed the fact of appraisal appears to be peculiar to certain types of media. A change in the type of medium implies a change in the type of appraisal and hence makes it difficult for one civilization to understand another (Innis 1950: 10).

Recalling Varela's idea of *enaction* (Sect. 6.3.1), it is not difficult to grasp the paradox of assessment's procedures which try to evaluate something *codifying it*, i.e., *producing it* (Benedikt's theory was mainly devised for designers).

This does not mean that such attempts are useless, indeed. Rather, it is a question of pursuing this goal being fully aware both of the enactive character of the research and of the irreducible chiasmus lying at the root of the issue at stake.

> If the identification of the real is based on the semantic univocality of the name, then the speech is unreal because equivocal, conversely, if we base the identification of the real on the semantic univocality of the proposition, then the nominal reference, as equivocal, it is unreal or relative to mere appearances (Melandri 1968: 217).

It is this the *chimeric* (logical and analogical) stuff maps are made of. A stuff better revealed by methods focusing on relations in themselves than on subjects already defined, like that of Simondon, as we have tried to elucidate. Even (or maybe above all) in a globalising world producing *hyperlocalism* (not only with regard to news coverage, but also to the production of hyperlocal networks), the interplay between places, information, identities, can hardly be disentangled without paying attention to such issues.

To tackle them a real transdisciplinarity is needed, as that pursued in the Batesonian 'metapattern' project (Volk 1995; Volk et al. 2007). Transdisciplinarity which, at least from a geographer's viewpoint, is a prominent resource in Simondon's works, as a promising key to the reconciliation of science and experience.

While it is true that physical place, social situations, cultures, are going to progressively separate, thanks above all to electronic media (first of all the Web), it also holds true that we still live in *places* not in *space*.

Space is an abstraction, sometimes useful, we seem to be accustomed to in order to reduce uncertainty in face of information surplus. Indeed, it is what mapping generates by blending logic and analogy, explicit and tacit, knowledge that and knowledge how.

Places, as many real experiences, are simpler *and* more complex than it is usually portrayed in the studies devoted to them. They are *simplex* (Berthoz 2012). So, if we really pay attention to Simondon's lesson, they will appear similar to the daring and endless filmmaking of *Othello* by Orson Welles, which hides, under the surface of outstanding settings and a wonderfully performed reading of Shakespeare's tragedy, a maze of places, times, characters, identities.

> Iago steps from the portico of a church in Torcello, an island in the Venetian lagoon, into a Portuguese cistern off the coast of Africa. He's across the world and moved between two continents in the middle of a single spoken phrase. That happened all the time. A Tuscan stairway and a Moorish battlement are both parts of, what in the film, is a single room. Roderigo kicks Cassio in Massaga and gets punched back in Orgete, a thousand miles away. Pieces were separated not just by plane trips, but by breaks in time. Nothing was in continuity. I had no script girl. There was no way for the jigsaw picture to be put together, except in my mind. Over a span of sometimes months, I had to keep all the details in my memory (Orson Welles, *Filming Othello*).

## References

Almeida I (2004) La boussole et la carte. Urbino: Quattroventi

Antonelli, C.: Localised Technological Change: Towards the Economics of Complexity. Routledge, London (2008)

Arendt, H.: The Human Condition. University of Chicago Press, Chicago (1958)

Arthur, W.B.: The Nature of Technology: What It Is and How It Evolves. Free, New York (2009)

Balibar, E.: Spinoza: from Individuality to Transindividuality. CIEPFC-Centre International d'Etude de la Philosophie Française Contemporaine. http://www.ciepfc.fr/spip.php?article236#nh38 (accessed 31 May 2013) (1993)

Balibar, E.: Europe as borderland. Environ. Plann. D Soc Space **27**(2), 190–215 (2009). doi:10.1068/d13008

Barabási, A.L.: Scale-free networks: A decade and beyond. Science **325**(412) (2009). doi:10.1126/science1173299

Bateson, G.: Mind and Nature. A Necessary Unity. Dutton, New York (1979)

Benedikt, M.: Information in space is the space in information. In: Michelson, Anders, Stjernfeld, Frederik (eds.) Billeder Fra Det/Images from Afar, pp. 161–172. Akademisk Volrag, Oslo (1996)

Benjamin, W.: Illuminations. Arendt, H. (ed.), translated by Zohn, H. New York: Harcourt Brace Jovanovich, paperback edition: 1969. Schocken Books, New York (1968)

Benkler, Y.: The Wealth of Networks: How Social Production Transforms Markets and Freedom. Yale University Press, New Haven [Conn.] (2006)

Berque, A.: Ecoumène: Introduction à l'Étude des Milieux Humains. Belin, Paris (2000)

Berthoz, A.: Espace perçu, espace vécu, espace conçu. In: Berthoz, A., Recht R (eds.) The Spaces of Man—Les Espaces de l'homme, pp. 127–160. O. Jacob, Paris (2005)

Berthoz, A.: Simplexity: Simplifying Principles for a Complex World. Yale University Press, New Haven, CT (2012)

Besse, J.-M.: Voir la Terre. Six Essais sur le Paysage et la Géographie. Actes Sud, Arles (2000)

Bowker, G.C., et al.: Toward information infrastructure studies: ways of knowing in a networked environment. In: Hunsinger, J., et al. (eds.) International Handbook of Internet Research, pp. 97–117. Springer, Dordrecht (2010)

Brose, H.-G.: An introduction towards a culture of non-simultaneity? Time Soc **13**(1), 5–26 (2004). doi:10.1177/0961463X04040740

Canguilhem, G.: Le vivant et son milieu. In: La Connaissance de la Vie. Vrin, Paris (English edition: 2001. The Living and its Milieu. *Grey Room.* Transl. By J. Savage. vol. 3, pp. 7–31) (1952)

Casey, E.S.: Representing Place: Landscape Painting and Maps. Univ of Minnesota Press, Minneapolis (2002)

Catoni, M.L.: La Comunicazione Non Verbale Nella Grecia Antica. Bollati Boringhieri, Turin (2008)

Cherubini, M.: Annotations of Maps in Collaborative Work at a Distance. Thèse École polytechnique fédérale de Lausanne EPFL n 4116. Lausanne: EPFL. http://dx.doi.org/10.5075/epfl-thesis-4116 (2008)

Ciolfi, L.: Augmented places: exploring human experience of technology at the boundary between physical and digital worlds. In: James-Chakraborty, K., Strumper-Krobb, S. (eds.) Crossing Borders: Space Beyond Disciplines, pp. 205–222. Peter Lang, Bern (2011)

Crampton, J.W.: Mapping: A Critical Introduction to Cartography and GIS. Wiley, Malden (2010)

Damasio, A.: Looking for Spinoza: Joy, Sorrow, and the Feeling Brain. Mariner Books, Orlando (2003)

de Boever, A., et al.: Gilbert Simondon: Being and Technology. Edinburgh University Press, Edinburgh (2012)

Dodge, M., Kitchin, Rob: Atlas of Cyberspace. Addison-Wesley, Harlow, UK (2001)

Doersch, C., et al.: What makes paris look like paris? ACM Trans Graph (SIGGRAPH 2012) **31** (3), 101 (2012)

Easley, D., Kleinberg, Jon: Networks, Crowds, and Markets: Reasoning about a Highly Connected World. Cambridge University Press, Cambridge (2010)

Farinelli, F.: I Segni del Mondo. Immagine Cartografica e Discorso Geografico in Età Moderna. La Nuova Italia, Firenze (1992)

Farinelli, F.: Geografia: Un'Introduzione ai Modelli del Mondo. Einaudi, Turin (2003)

Farinelli, F.: De la Raison Cartographique. CTHS-Éditions, Paris (2009)

Ferraris, M.: Documentalità. Perché è Necessario Lasciar Tracce. Laterza, Rome (2009)

Galison, P.: Einstein's Clocks and Poincare's Maps: Empires of Time. W.W. Norton Company, New York (2004)

Headrick, D.R.: When Information Came of Age: Technologies of Knowledge in the Age of Reason and Revolution 1700-1850. Oxford University Press, New York (2000)

Hendler, J., et al.: Web science: an interdisciplinary approach to understanding the web. Commun. ACM. **51**(7), 60–69. doi:10.1145/1364782.1364798. http://doi.acm.org/10.1145/1364782.1364798 (2008)

Holenstein, E.: Philosophie-Atlas : Orte und Wege des Denkens. Ammann, Zürich (2004)

Innis, H.A.: Empire and Communications. Oxford University Press, Oxford (1950)

Kapp, E.: Grundlinien einer Philosophie der Technik. George Westermann, Braunschweig. (French edition: 2007. Chamayou, G. (ed.): Principes d'une philosophie de la technique. Vrin, Paris (1877)

Kavouras, M., Kokla, Margarita: Theories of Geographic Concepts: Ontological Approaches to Semantic Integration. CRC, Boca Raton (2008)
Kitchin, R., Dodge, Martin: Rethinking maps. Prog. Hum. Geogr. **31**(3), 1–14 (2007)
Koselleck, R.: The Practice of Conceptual History: Timing History, Spacing Concepts. Stanford University Press, Stanford (2002)
Iacono, A.M.: Autonomia, Potere, Minorità. Del Sospetto, della Paura, della Meraviglia, del Guardare con Altri Occhi. Feltrinelli, Milan (2000)
Leroi-Gourhan, A.: Le Geste et la Parole: la Mémoire et les Rythmes. Albin Michel, Paris [English edition: Leroi-Gourhan, A. 1993. Gesture and Speech. Cambridge (Mass.): MIT Press] (1965)
Libet, B.: Mind Time: the Temporal Factor in Consciousness. Harvard University Press, Cambridge, MA (2005)
Lima, M.: Visual Complexity: Mapping Patterns of Information. Princeton Architectural, New York (2011)
Livingstone, D.N.: Putting Science in its Place: Geographies of Scientific Knowledge. University of Chicago Press, Chicago-London (2003)
Melandri, E.: La linea e il circolo. il Mulino, Bologna (1968)
Meyrowitz, J.: No Sense of Place: the Impact of Electronic Media on Social Behavior. Oxford University Press, Oxford (1985)
Miconi, A.: Reti. Origini e Struttura della Network Society. Laterza, Roma-Bari (2011)
Morelli, U.: Mente e Paesaggio. Una Teoria della Vivibilità. Bollati Boringhieri, Turin (2011)
Musil, R.: Denkmale [Monuments]. In: Frisé, A. (hrsg.) Gesammelte Werke. Rowohlt, Reinbek bei Hamburg, VII (English edition: 1987. Posthumous papers of a living author. translated by Wortsman, P., Hygiene. Eridanos, Colo) (1978)
Neve, M.: Virtus Loci. Lineamenti Fondamentali di una Teoria dell'Informazione Spaziale. Quattroventi, Urbino (1999)
Neve, M.: Il remo e lo stampo. In: Barbanti, R., Boi, L., Neve, M. (eds.) Paesaggi della Complessità. La Trama delle Cose e Gli Intrecci Tra Natura e Cultura, pp. 283–315. Mimesis, Milan (2011)
Polanyi, M.: Personal Knowledge. Towards a Post-Critical Philosophy. Routledge and Kegan Paul, London (1958)
Raffestin, C., Bresso, Mercedes: Tradition, Modernité, Territorialié. Cahiers Géographie Québec **26**(68), 185–198 (1982)
Robinson, A.H.: It was the mapmakers who really discovered America. Cartographica **29**(2), 31–36 (1992)
Roy, O.: Holy Ignorance: When Religion and Culture Part Ways. Columbia University Press, New York (2011)
Russell, D.: Open Letter on the Dematerialization of the Geographic Object. In: Gale, Stephen, Olsson, Gunnar (eds.) Philosophy in Geography, pp. 329–344. D. Reidel, Dordrecht (1979)
Shadbolt, N., Berners-Lee, T.: Web science emerges. Sci Am **299**(4), 76–81 (2008). doi:10.1038/scientificamerican1008-76
Simondon, G.: Du mode d'existence des objets techniques (nouv. éd. 1989). Aubier, Paris (English unpublished edition: 1980. On the Mode of Existence of Technical Objects, transl. by Ninian Mellamphy with a Preface by John Hart. London, On.: University of Western Ontario) (1958)
Simondon, G.: L'individu et sa Genèse Physico-Biologique Individuation à la Lumière des Notions de Forme et d'Information. PUF, Paris (new edition J. Millon, Grenoble, 1995 ) (1964)
Simondon, G.: L'Invention dans les Techniques: Cours et Conférences. Édition établie et présentée par Jean-Yves Chateau. Éd. du Seuil, Paris (2005)
Simondon, G.: Cours sur la Perception (1964–1965). Les Éditions de la Transparence, Chatou (2006a)
Simondon, G.: Imagination et Invention (1965–1966). Les Éditions de la Transparence, Chatou (2008)
Simondon, G.: Communication et Information. Cours et Conférences. Les Éditions de la Transparence, Chatou (2010)

Star, S.L., Ruhleder, Karen: Steps toward an ecology of infrastructure: design and access for large information spaces. Inf Syst Res **7**(1), 111–134 (1996)

Star, S.L.: This is not a boundary object: reflections on the origin of a concept. Sci Technol Human Values **35**(5), 601–617 (2010). doi:10.1177/0162243910377624

Stone, L., Lurquin, P.F., Cavalli-Sforza, L.L.: Genes, Culture, and Human Evolution: a Synthesis. Wiley, Malden, MA (2006)

Tagliagambe, S.: Lo Spazio Intermedio. Rete, Individuo e Comunità. EGEA—Università Bocconi Editore, Milan (2008)

Tagliagambe, S.: Città e spazio pubblico. Organizzazione delle reti e nuove conoscenze. In: Democrazia. Storia e teoria di un'esperienza filosofica e politica, C. Altini (a cura di), 413–447. il Mulino, Bologna (2011)

Thrift, A.: Spatial Formations. Sage, London (1996)

Tuan, Y.-F.: Space and Place. The Perspective of Experience. University of Minnesota Press, Minneapolis (9th printing 2002) (1977)

Tuan, Y.-F.: The significance of the artifact. Geogr Rev **70**(4), 462–472 (1980)

Turnbull, D.: The ad hoc collective work of building gothic cathedrals with templates, string, and geometry. Sci Technol Human Values **18**(3), 315–340 (1993)

Tversky, B.: La cognition spatiale: incarnée et désincarnée. In: Berthoz, A., Recht, R. (eds.) The Spaces of Man—Les Espaces de l'homme, pp. 161–184. O. Jacob, Paris (2005)

Varela, F.J., Thompson, E.T., Rosch, E.: The Embodied Mind: Cognitive Science and Human Experience. The MIT Press, Cambridge, MA (1993)

Virno, P.: Moltitudine e individuazione. In: Virno, P. (ed.) L'individuazione psichica collettiva, pp. 231–241. DeriveApprodi, Rome (2001)

Virno, P.: E Così via, All'Infinito Logica e Antropologia. Bollati Boringhieri, Turin (2010)

Volk, T.: Metapatterns: Across Space, Time, and Mind. Columbia University Press, New York (1995)

Volk, T., Bloom, J.W., Richards, J.: Toward a science of metapatterns: building upon Bateson's foundation. Kybernetes **36**(7/8), 1070–1080 (2007). doi:10.1108/03684920710777847

Wenger, E.: Communities of Practice: Learning, Meaning, and Identity. Cambridge University Press, Cambridge (1999)

Wood, D.: The anthropology of cartography. In: Roberts, Les (ed.) Mapping Cultures: Place, Practice, Performance, pp. 280–303. Palgrave Macmillan, Basingstoke (Hampshire) (2012)

Wood, D., Fels, J.: The Natures of Maps: Cartographic Constructions of the Natural World. University of Chicago Press, Chicago (2008)

# Chapter 7
# On Growth and Form of Narrative Structures

**Guido Ferraro**

**Abstract** Narrative is firstly a formal organization, but it is a form that interprets the events giving them meaning. Starting from a reinterpretation of the classical *Morphology of the Folktale* by Vladimir Propp, we can now note how narrative form and cultural meanings interact with each other. Thus, we remove the "formal" dimension from its traditional segregation to a universe of insubstantial non-things, returning it to the arena of human strategic action and social practices. We may conceive a story as a route performed by a subject on a social and categorical map: so, a narrative configuration is essentially a set of dynamic relations, lying between a procedural and a systemic dimension. We find the basis of everything in the fundamental Saussurean view that interrupts the ordinary separation between "things" and "relations": identity, meaning, and structure are the effects of systemic relations. In this light, Claude Lévi-Strauss offers us the most elaborate picture of narrative systems, where textual objects are seen as secondary outcomes of transformational tensions: every text is by nature a remake; it exists only through other texts. The most radical feature of this original perspective is indisputably the adoption of the theoretical model expounded by D'Arcy Wentworth Thompson in his famous book on the morphogenesis of zoological species, *On Growth and Form.* Textual theory emerges greatly innovated, linked to a view of cultures as systemic networks of connected texts. And this applies also to products of our culture, as the concluding examples (the *Alien* film saga and Puccini's *Bohème*) should positively illustrate.

## 7.1 The Meaning of the Form

For some time, the idea that the narrative system has a central or even primary role among other semiotic systems is widespread, not only in semiotics but also in psychology and other areas of the humanities. Take, for example, the opening

G. Ferraro (✉)
Università degli Studi di Torino, Torino, Italy
e-mail: guidfer@gmail.com



A. Sarti et al. (eds.), *Morphogenesis and Individuation*,
Lecture Notes in Morphogenesis, DOI 10.1007/978-3-319-05101-7_7

words of Hayden White's famous book *The Content of the Form*, which is undoubtedly relevant to our discussion: "To raise the question of the nature of narrative is to invite reflection on the very nature of culture and, possibly, even on the nature of humanity itself" (White 1987: 1). This primary and dominant position —that could for the first time free language from the role of "primary modeling system"—seems attributable to the fact that the narrative system offers us an image of special abstraction, or of particular purity, where the form dominates what represents the filling, i.e. the concrete elements of the story. The distinction in the field of historiography between real history and simple chronicles is revealing. "Form" is the manner, imbued with narrative, in which historians organize the data provided by the material in the chronicles. This form gives meaning to events that, by themselves, do not appear to have a narrative configuration. As White writes (1987: 44), "any given set of real events can be emplotted in a number of ways, can bear the weight of being told as any number of different kinds of stories". So, "it is the choice of the story type and its imposition upon the events that endow them with meaning". The narrative, in short, adds nothing to the content of the events that it tells, but superimposes a formal organization, and since this form interprets the events, it is crucial to their meaning. Let's start from the idea that the form is anything but a neutral instrument of meaning, and let's discard from the very beginning the hypothesis that sees the form that organizes a story as a mere reproduction of the "form of the events" to which it alludes. Multiple hypotheses are still available, which are very different from each other.

The book to which we refer most often, as the starting point of narrative theory, holds the significant title of *Morphology of the Folktale* (Propp 1928). This work primarily presents an "unitary composition scheme", intended to describe the general form of folktales from the Russian tradition. The scheme, despite being constructed in a partially inductive manner (from one hundred textual samples), can neither be described as a simple generalization nor as a real prototype. Since no folktale presents the entire range of functions that make up the whole schema— some, in fact, present very few, or select a conspicuously partial subset—you could say on one hand that the composition scheme is, as such, missing direct textual implementation, and on the other that each story is seen as an imperfect trace of the model, as an incomplete and basically inessential occurrence. What matters are not the projections of the model onto textual objects, but the semio-cultural configuration as such: an higher-level entity, not directly observable, understandable only in its whole finished design, and as such the primary carrier of semantic values. This hypothesis is perhaps questionable, but intriguing.

The cultural value of this configuration was subsequently clarified by an analysis of its *historical roots* (Propp 1946), which is founded in the fundamental ritual practice of initiation ceremonies. This historical and cultural grounding, badly lost in subsequent generalizations, should make us aware of the fact that the schema matches one of the different existing narrative architectures. In any case, this inaugural essay, permeated by the spirit of the formalist movement, has bequeathed us with the idea that it is to some extent possible to isolate a purely formal level, provided with relative autonomy.

For its part, the subsequent Greimasian school, despite having insisted on the concept of "immanence", nevertheless allocated immanent nature only to the variable textual content, and certainly not to the formal structures called to organize them—regardless of whether we are referring to the "semiotic square" proposed to structure the deep level or the "canonical schema" which sets up the surface narrative structure. With respect to the first structure, we return to a logical dimension found in the most extreme levels of generality and abstraction, trying even the dubious move of an Aristotelian complicity. The second structure invokes the interesting but imprecise concept of an analogy of life experience. Neither these, nor other formal structures proposed in the course of the volcanic wave of Greimasian innovations, are marked by historic value, and therefore meaning. In the case of the semiotic square, the meaning comes from the elements called to fill the slots, not from the logical form, conceived as unchanging and neutral. The same can be said for the "canonical narrative schema", which in spite of it's limited starting textual reference was immediately, carelessly generalized, taken on as an universal model of narrative construction. The reference to the formal mechanism does not imply, in this context, a choice, axiological implications, or semiotic functionality. Regardless of whether they are logical structures or experiential patterns, they are projected onto the text as if they were plummeting down from extra-semiotic spaces. One may well ask at this point if other avenues are possible, theoretically better developed and more suitable for insertion into current perspectives. As we will see, in this sense important traces in the history of semiotics are identifiable. One must, first of all, not think of "form" as a kind of statically defined mold or matrix, located on an uneven plane compared with that of textual objects, and therefore removed from the mechanisms of the "framework of social life", which Saussure tied to his original conception of semiotics. One must also remove the "formal" dimension from its traditional segregation to a universe of non-things, as impeccable as insubstantial, in order to return it to the arena of human action, endowed with meaning and strategic direction, in the middle of "social practices". As we shall see, there are semiotic theories that move in this direction. However, from the beginning we must remember the fundamental Saussurean concept that interrupts the ordinary separation between "things" and "relations". He claims that the identity of things is definable in terms of a pure set of relationships, without requiring full entities, or "positive terms" with which to engage for their institution. But if we really want to understand in what sense it is possible to identify a formal component, independent or otherwise, we need to clarify something regarding the basic conditions of narrative.

## 7.2 The Essence of Narrative Form

The concept that narrative constructions are based on an elementary structure is largely accepted. Gerald Prince (1982), who addressed this in more detail than others, called it "minimal story". It is composed of three segments: A. a departure state, B. a transformative event, C. an arrival state. Between the first and the second segment there is a simple temporal succession, while between the second and third, there is also a relationship of causal determination. Finally, the state of things C, presented in closure, should not be thought of as independent, and must correspond to a possible transformation of the departure state A. Let us add that this also assures the hold of the thematic continuity. If, for example, the opening segment thematizes the economic status of a character X, that we assume suffers from poverty, the closure state should also address the relationship between X and an economic state. In the C phase he could have become enormously rich, simply not as poor, or have gotten even poorer. This, in any case, must appear as a result of the way in which the event B has transformed the initial state A.

Of course, most stories are far more complex. But the first explanation for this is that elementary structures of this type are simply joined in sequence, so that the story presents itself as a set of segments, in which each segment is defined by one entry and one exit configuration. Another interesting mechanism is one in which a minimum story (described earlier) generates a series of expansions that can even create a very long and complex construction. For example, the starting condition, "X is poor", can be expanded into a sub-narration that explains how X has been impoverished, say, losing his possessions at poker. In a next step, the transformative event "losing his possessions at poker" can be expanded into an episode with greater detail, which shows how his dearest friend transformed him from a prudent father into a reckless gambler. It can continue in this manner theoretically to infinity. At each step, what was a single phase of a simpler structure gives rise to an entire sequence. The mechanism is elementary, if you will, but offers us a way of looking at *text generation* in a very different manner from the Greimasian "generative trajectory", often criticized for its difficulty in explaining *conversions* between levels that are too markedly heterogeneous. The generation of a story does not start with an abstract logical structure, but from a basic narrative structure. In addition, this perspective offers us a substantial isomorphism between the overall shape of the story and the configuration of the segments that compose it. At all levels, the main idea remains the same. A beginning condition is given, defined as relational configuration between a number of elements. A transformative process, changing the relationships between those elements, establishes the end condition. The vicissitude that corresponds to the classical Propp-Greimas pattern shows, for example, the transformation of a character's identity, originally in a state of social marginality, later achieving a status characterized by wealth and power.

The concept of a *transformative event* corresponds in substance to the Proppian concept of "function". Although the original definition was formulated in an unsatisfactory manner, we can utilize the suggestion related to the use of a term that, at least in Western languages, has a key role in both the mathematical and computer theory. The latter is particularly interesting, in that in computer theory a strictly defined procedure, which outputs the result of a transformation of the input terms, is called a "function". The analogy is therefore not superficial. For example, when we speak of the function called 'Attainment of the magical tool', we refer to an incident that leads to a condition defined in relation to the departure point—the

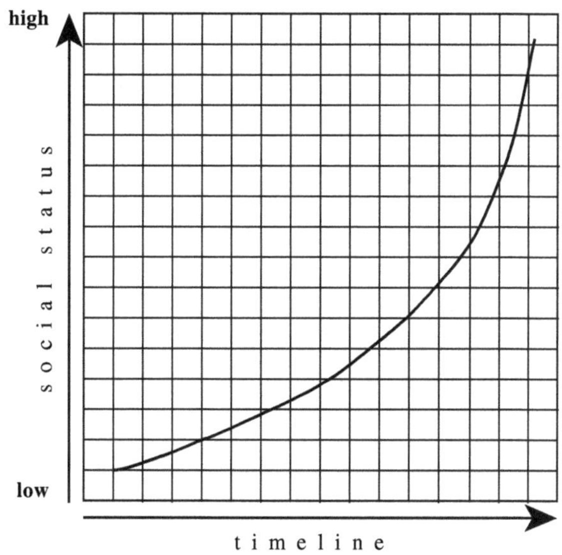


**Fig. 7.1** The grid represents the invariable social system, while the *curved line* represents the individual route: a change of position in the grid

protagonist must do something but lacks one or more essential tools—into a different output condition. In this case, the hero now has the tools needed to accomplish the feat. With each function that happens in the story, the situation—that is, the relationship between the elements at hand—changes in a certain way, corresponding to the transformative nature of that particular function. Even if in the various narratives the elements may be completely different, the function processes this relational transformation following the same logic. Therefore from Propp's research we derive the idea that the transformative phases, central to any narrative construction, do not require a local description, but correspond to a limited set of procedures that are grammaticalized on a, in some way, global level. On top of this, there is the fundamental concept that a narrative model (or "compositional scheme", in Proppian terminology) corresponds to an essentially fixed string of functions. The elements that are introduced have little importance, since the meaning comes mainly from the global design of this transformative mechanism.

From this basic theoretical model, it is possible to add at least two more elaborate, different but not mutually exclusive, notions. The first notion, which examines the relationship between the departure and arrival states, recognizes the presence of a fundamental hierarchical relationship that definitely favors the second state with respect to the first. Greimas's definitions are the most well known in this regard. He refers respectively to *placed content* and *reversed content*, but often this view is not only connected to, but confused with Lévi-Strauss' concept, from which it was originally inspired. Both authors believe that the situation presented in the final stages of the story is that which is positively supported and affirmed by the text, and which is reversed in the initial situation. There are, however, decisive differences. In Greimas' version, everything is tied to the perspective of a subject

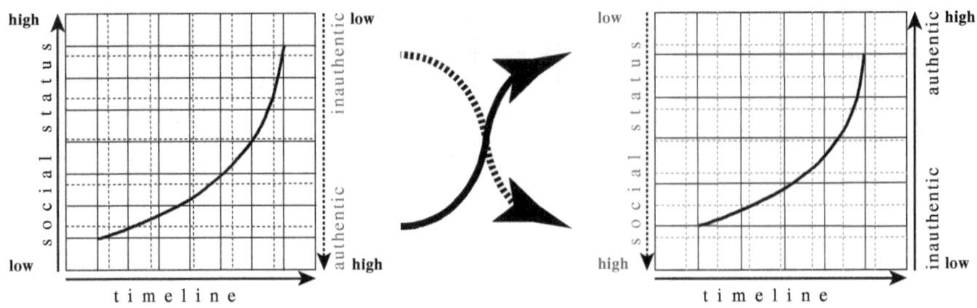


**Fig. 7.2** Here the individual route remains unchanged, but the subject decides to change the coordinates defining the frame of reference he is using to read the reality. So, the value of the opposite poles turns out to be reversed, in both couples concerning social status and authenticity. Same route, but in an altered grid

that formulates a narrative program that will lead to the final condition, which is initially placed in the dimension of desire and virtuality. In Lévi-Strauss' case the final state, on the contrary, far from being marked by virtuality, corresponds to the state of things as we can observe them in the world around us, while what is projected in the past tense ("mythical times"), therefore placed in the initial phase of the story, is its hypothetical, virtual conceptual variation. The theoretical question—for instance, around the value of the seasons or the meaning of funeral rituals—is answered through the examination of an alternative condition: "And if the seasons did not exist? And if funeral rites were not held?" In both cases, however, the beginning and end no longer appear to us as simple poles of a linear sequence, but rather as hierarchized constructs, dispersed on the timeline for fundamentally semantic reasons. We can already say that the story, rather than "moving" syntagmatically from one extreme to another, compares paradigmatically alternative conditions.

However, another direction that we can examine opens up, as stated earlier. The narrative arc, which we superficially see as a bridge between the initial and final state, establishes a *polarity* that separates, and somehow keeps the two conditions at a distance. Obviously, since all of the elaborations of the basic formula are possible, we can encounter every imaginable variation, including those of narrative texts in which the final state is no different from the initial one. We know, however, that the basic form tends to mark a clear difference (for example, the character that will eventually be rich and powerful is poor and marginal at the beginning). In the framework of the creation of an integrated semiotics[1]—what I call "neoclassical" semiotics—this differential tension was linked to (quite spontaneous for those who work in a Saussurean manner) the concept of *difference* as the primary root of signification. It is quite logical to consider the narrative construction as a device that manages the projection (stretched across the expositive dimension of the sequence) of a semantic value, by its nature defined precisely in terms of a difference. The typical narrative path, that runs from the disjunction from some object of value

[1] Ferraro (2013): 183–192.

(taken as the initial condition), to the conjunction as a final condition, may well be rethought of as the result of the syntagmatic spreading of a value construction. Contrasting absence with the presence of a marked entity, its value becomes evident. In this manner, for example, by showing the difficulty, the misery and sorrow of those who live under foreign rule in the beginning, and showing the satisfaction, the benefits and happiness of those who enjoy freedom and independence in their own homeland in the end, one could express a thorough semantic investigation of the concept of "national independence", pathemic components included. The narrative results in an exploration of an area of the cultural system.

In the neo-classical semiotic project, aimed at overcoming the traditional separation between semiotic research areas, this analytical method represents a pivoting point, as it allows a decisive structural connection between narrative theory, sign theory, and the theory of passions. Even the latter, in fact, responds to the same construction principle. What we call "passion" is the effect of meaning resulting from the comparison between two alternative states of affairs (the comparison between the possessing or not possessing a certain thing represents, of course, only one of many discernable cases). In short, we have at our disposal the first draft of a basic form that, by developing the Proppian concept of function, brings many key areas of semiotic theory closer to each other. An essential concept lies at the root: what appears to have a positive identity should instead be conceived of as a basically relational structure.

## 7.3 The Process and the System

This "nuclear" conception of the story, while presenting many interesting aspects, as we have seen, is an overly simplified schematization. In the context of this article, centered on the genesis of narrative forms, it should be especially emphasized that a redefinition of Proppian "functions" in terms of transformative procedures neglects the important question of the authority to which these regulatory mechanisms refer (therefore, in Greimasian terms this concerns the role of the Sender, with all its variants and its rightful extensions[2]). For example, consider the aforementioned segment of the Proppian compositional scheme in which the subject of an action finds themself unable to carry out their narrative plan for lack of knowledge, skills, tools, or transportation. Citing the transformative process—which will change the relationships between the elements at play, and output a subject prepared to accomplish the task—is not enough here, we must also specify that this procedure is initiated and controlled by a defined top-level entity (according to Propp, this corresponds to the Donor, and for a more elaborate narrative semiotics to the Sender of use values).

This means that the *function* status is more complex, because it must take into account both the completed transformation (or the transformation that is planned,

[2] See Ferraro (2012).

expected, and hoped to be realized) and the authority that presides over it as well as the control logic that it applies. These forces, that design and govern the transformation procedures, are essential in determining the meaning of the events. That Romeo and Juliet could shift from being simple lovers to legitimate spouses (Wedding function) is certainly interesting, but it is obvious that the meaning becomes very different if they marry by choice and without difficulty, if they are forced to do so by their parents, or if they marry with the complicity of mothers but against the wishes of their respective clans, and so on. Every solution bears very different meanings, precisely because each of these possibilities places the transformative action in a different position within the *map* provided by the normative framework.

This concern for *rules* is consistently present in the narrative constructions. Even if we look only at Propp's compositional scheme, we can see that the function couple Prohibition/Infringement is proposed from the very beginning. In as far as the value assigned to these functions can be significantly variable (the prohibition may be proposed with correctly protective intent, or conversely as unlawful limitation of the protagonist's free will), we should still take note that, from the beginning, a dialectic is proposed between dependence and autonomy, and with this an unequal condition of knowledge and authority, and an opposition between rules coming from the outside and the choices made by the protagonist. The first indication of what we call the "Subject" is presented in terms of an ability to define routes and plans of action, coming out of a "home" to be understood not only as a protective family but also as a reference to the systemic dimension of shared norms. This narrative architecture immediately declares that the subject does not exist if not through its own distance from the plane of constructed models. This is immediately seen as an irrational and self-destructive gesture. The story may later lead the protagonist to a positive redefinition of their identity and their relationship to the surrounding universe. However, *going through the forest* despite mother's prohibition, as happens in many fairy tales, appears at first as silly as Ulysses' decision to go and see what there was beyond the pillars of Hercules was traditionally considered "crazy".

In more general terms, we can see that much narrative architecture is built on the correlation of two planes, respectively based on profoundly different logic. One plane is based on what I call the Perspective Rule, transformative and innovative, moved by the desire of a Subject that intends to achieve valued goals (typically, acquire social status). On the other hand there is a Destination Rule, called on to uphold the syntactic positions that belong to various types of Senders, but which correspond first of all to a shared plane of norms and patterns. If the first Rule is activated by the subject's willingness to change its position in the social system, the second is instead a systemic tendency towards homeostasis and the reconstitution of balance. Each of the two planes has its own distinct functional logic and its own manner of story construction, as well as its own evaluative logic for the assignment of meaning to things. These evaluation methods have a *subjective* nature for the Perspective level and an *objective* nature for the Destination level. Hence the definition of two distinct sets of semantic assignments, and the construction of separate Objects of value. The architecture that best shows this duality is that which is based on the Contract, because by definition it implies an exchange of Objects of value, and on a deeper and interesting level, the possibility for *translatability* between objective and subjective values. In this manner, for example, the killing of a dragon that threatens the community (somatic act on the objective level) can be translated, on the subjective level, into the decisive transformation of a strictly personal identity.

Precisely as the contract-based architecture shows us, the two *procedural* and *systemic* narrative levels in fact implicate each other. We see that the social authority (in the fairy tale, typically, the king) is not able to act on the objective plane in the absence of a subject who is looking for an identity. On the other hand, this subject cannot hope to achieve their desired identity until some urgency is objectively reported. This relationship of implication can be made equally clear even in a completely different architecture. For example, when the subjective and private love affair between characters like Romeo and Juliet, otherwise devoid of particular interest, collides with a conflictually structured social system.

As one can imagine, a significant portion of the meaning in a narrative text (or in the model that underlies it) lies in the way this fundamental relation is defined, demonstrating if the subject asserts itself through the observance of the rules, or conversely rebelling against them. A factor that has decidedly limited, and somehow diverted the development of narrative theory, is the obsessive reference to an architecture, such as that of typical Russian fairy tales, that is characterized by a marked respect for authority, and therefore by a principle in which the definition of personal identity comes from above, provided through strongly coded elements (noble status, accentuated levels of wealth, etc.…). In this case a kind of static map of the world predominates. The subject, even with all of its agency, has no intention of changing this map. Instead, the subject has every interest in keeping the structure firm, simply changing its place on the map. The story never questions the grid on which is drawn, but merely indicates the distance between the starting and ending position of the subject. The final function that Propp astutely called Wedding (even without fully grasping its value) is particularly emblematic in this sense. The Wedding, more than just a positive sanction, overlaps a prize value with the very interesting concept of a merger. The Subject, initially very far from the heart of the social mechanism, affirms himself by dissolving into the authority, marrying and becoming King himself, a guiding principle, foundation of that unalterable world map on whose entanglements he earlier actively climbed (Fig. 7.1).

## 7.4 Questioning the Frame of Reference

We need not think about the new world that opens up beyond the Pillars of Hercules in order to explore a very different architecture than that which Propp studied. A mundane transatlantic cruise could illustrate how a modest route can be equally significant (it is symptomatic that the representation makes use, in any case, of a spatial distribution). I'm referring to the story told in the film *Titanic* and to the small vertical journey that leads a girl, initially decked in first class with her mother and future husband, to descend into the far more lively and intriguing third-class deck, falling into the arms of a romantically creative but penniless passenger. It is evident here how different the pillars that hold up the narrative architecture are. The girl is very aware that she is not moving on a static social and categorical map. With her gesture, she is affirming another way to: read the world; determine the relationships between the values; conceive individual destiny; evaluate the impact of the rules. All of this is irreducible to the reading of world from which she comes. In particular, it becomes clear that individual destiny is no longer the variable element in a fixed reference system, since what is questioned in the first place is the reference system itself. In a certain manner, the configuration of the linear development does not fundamentally change. The beautiful Rose must still identify the person with whom she feels an affinity strong enough to impel her to join their life. The development of events remains substantially the same (the girl does not decide, say, to give up a betrothal in order to instead enroll in a university course, but rather inhabits the same type of story with another logic), the difference lies in the shift of interested values—those that in our metaphorical use of the Cartesian model would be found on the ordinate axis. The story's audience clearly perceives that the change of reference values makes the girl protagonist similar to the romantic artist in third class, and hopelessly dissimilar to the businessman boyfriend with whom she otherwise shared the opulent spaces of first class. Maintaining the route taken by the girl substantially stationary only serves to emphasize the comparison and difference between the reference systems, and makes it clear that this is precisely what matters (Fig. 7.2).

Far from being an anomaly, this example indicates how limiting the flattening of the story to a mere linear process of transformation can be. In fact, quite central is the relationship that situates the transformative process in the design of the very space that maps the possibilities, the categorical axes, the reference points, and ultimately the criteria for creating meaning. The model of the map and route, or in my terms the intersection between the two fundamental principles of Destination and Perspective, can be seen at the root of many narrative phenomena, as well as at the root of ways of building sense in many semiotic systems. I will cite just one example, whose distance provides an idea of the conceptual model's range. According to Byron Almén (2008) music, in its various forms, has a construction similar to that of the narrative. Each piece of music is based on a tension between order and chaos. On one hand, it provides a system of tonal and formal rules, but on the other hand, the composition's development derives its appeal from the fact that in some way it seems to question the system itself. The musical phrases, in a certain sense equipped with a kind of agentive force, present dissonances and irregularities, delays and detours. These introduce conflict, the possibility to disrupt order, a virtual effect of chaos. The musical architecture is therefore based on a conflict between a hierarchy and a transformation principle, whose results may be different depending on the case. Will the order be maintained in the end, or will new and unexpected paths open? Will the hierarchical system be upset by the transgressions, or will a synthesis of initially conflictual components be reached? We are faced, once again, with a variant of the model proposed earlier. We can understand that the confirmations could be multiplied, utilizing reflections from many different semiotic fields. But at this point it is sufficiently clear in what sense the narrative does not propose a linear process, but rather a complex and ever-changing relationship between the dimension of the process and its systemic context, constantly called into question, or if we prefer, between the journey and the map on which it takes place. Now we can use a theoretical conception that develops these ideas, reorganizing them in an original and advanced direction.

## 7.5 The Concept of "Transformation", from Thompson to Lévi-Strauss

Claude Lévi-Strauss, one of the most important masters of narrative theory, proposed a very different perspective on the generation of stories from those most common in semiotics.[3] He does not refer to the generation of text from an external cultural model (as in practice we find in Propp), nor to its generation from a deep internal core (as in the Greimasian path). His concept is radically structural and Saussurean. As the Genevan teacher claimed, semiotic objects are presented as a kind of epiphenomenon that veils their true nature as essentially relational entities. This principle is applied by Lévi-Strauss both to the specific symbolic components present in stories and to whole texts and mythological complexes. The concept of "relational identity" is methodologically rendered through the introduction of a completely new notion of "transformation".

In this perspective, the narrative heritage of a culture is no longer thought of as a set of stationary and closed objects, with independent and internally defined identity (that which we usually call "texts"), but rather appears in the form of an uninterrupted flow of ordered operations, those for which we would now use the term *practices*. These continually reinterpret and rewrite the myths, so that it is never possible to identify an "original" or "definitive" text. Each story is nothing but the coagulation of a transient state of the symbolic system. On the other hand, the system only manifests itself through its precarious projection onto a series of texts. If what matters the most in every story is the differential range that defines and identifies it, the "structure" does not seem as important as its transformative capacity. In this view, the texts acquire meaning through the interweaving of their mutual references, by playing with analogy and reversal, in a fascinating mechanism of incorporation and endless *remaking*. Every text, one can say, exists only *through other texts*. Every myth is by nature a translation, deriving its origin from another myth in a neighboring but foreign population, or from a myth in the same population, temporally earlier or contemporary, or perhaps belonging to another social division. Far from being mere formal operations, these practices visualize the

[3] For an overall reappraisal of Levi-Strauss' narratological theory, see Ferraro (2001).

signs of tensions that, even in the simplest ethnological realities (we will shortly see an example), create conflict between social forces, cultural models or ethnic groups. This demonstrates how Lévi-Straussian theory offers valuable elements for those who intend to propose seriously *socio-semiotic* perspectives.

Lévi-Strauss needed a "transformation" concept that involved more than the comparison and analysis of textual divergences, beyond the line of succession or the diachronic dimension. The "transformations" to which he refers do not lie between a "before" and "after", but rather between the one and the other's *systemic contexts*. He found the most appropriate theoretical model to build on in the famous book on the morphogenesis of zoological species, *On Growth and Form*, by D'Arcy Wentworth Thompson (1952). From this book, great in every sense (the first edition is 793 pages, 1116 the second), Lévi-Strauss essentially used the chapter on the *Theory of transformations*, which is also the best-known part in general.

Thompson's basic idea is that—in the study of the morphology of an animal species, in his case—we should take greater account of the integrity of the whole, instead of examining the parts as if they were independent components, a live body is indivisible. The goal, therefore, is to render scientifically evident the way in which living forms are related to one another as a whole, considering the observable differences between related species not in terms of a sum of individual modifications but as a process that informs the integral unit. For us, the decisive points are twofold: that the differences between entities of any type can in fact be expressed in a unified form, as the result of a single overall transformation which gives rise to a whole set of related changes; and that these processes entail the action of defined dynamic forces. Let us add that these "transformations", according to Thompson, are not temporally oriented. In other words, they do not imply that a species is descended from another, or that one species is the result of a Darwinian modification of another. The transformative is by principle a reciprocal relationship, which can be thought of in one direction as much as in the opposite one. Lévi-Strauss uses precisely these concepts, but also ideally incorporates the use that Thompson had made of topological models. In order to create a visible and precise representation of his way of thinking, he proposed that the relations between similar species could usefully be displayed on a Cartesian coordinate system. Here you find, in fact, the more evident passage from the concept of a set of biological components to a properly systemic entity. Thompson shows (see the examples shown in Fig. 7.3) how the differences that seem ambiguous and complex can be instead referenced to simple phenomena, assuming you shift the focus from the objects to the coordinate system in which they are placed, and from which they are (mathematically, in this case) defined. The deformation of the coordinate system explains the consequent alteration of what is inscribed, so that a series of changes that we would have first identified at the level of the specific components becomes, in fact, aspects of a single overall transformation.

Whatever value that Thompson's theory still has in the biology, I think we can say that the re-use of these concepts in the human sciences could have an even greater scope and significance. Topological models developed in mathematics come to anthropological studies, semiotics and the theory of narrative, through the helpful

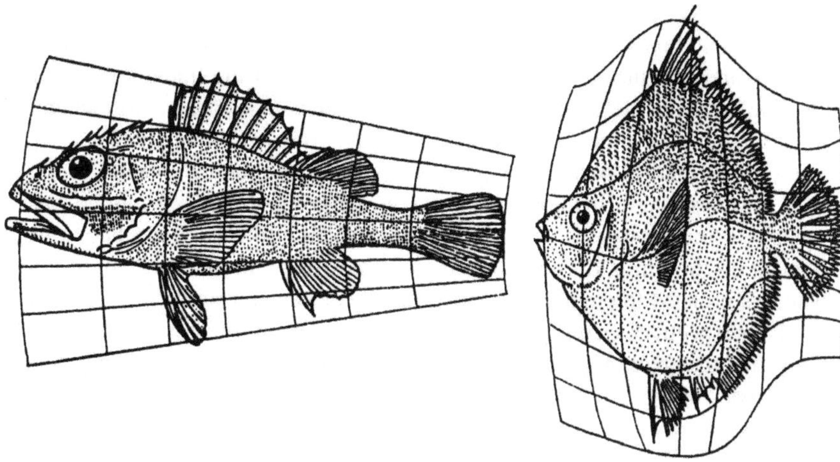

**Fig. 7.3** A typical example from D.W. Thompson's *On Growth and Form*, p. 1063

filter of biology. Worth thinking about, in any case, is the fact that Thompson's conceptual innovation received much attention in the field of biological sciences, while it has gone largely unnoticed, and therefore unused in the field of semiotic research (perhaps more short-sighted, tending to be more dogmatic). The idea that the differences between observable objects (that is, in our case, between texts) can be much better defined and explained by returning them to the differences between the *reference systems*, is of fundamental interest, even if in this case we mean semiotic systems, not subject to mathematical descriptions and only metaphorically close to Cartesian systems.

Even here, the differences between the stories are interpreted as effects of the transformation of the containing field, or as a passage of the textual object through different containing fields, as we will see shortly in some examples. We must also stress in this respect (even though we cannot further develop this discourse here), that the shift of attention from objects to the system in which they exist also corresponds to a reintroduction of a Saussurean vision (accent on system rather than on the textual product) in a more effective and mature manner than either the excessively simple original, or the equally excessive shift to textual perspectives. The conceptual distinction between the coordinate system and the path that is inscribed inside it remains decisive, even if abstracted from its mathematical reference, confirming in another way the presence of two basic components: a generally static regulatory level and a dynamic and transformative procedural level.

It goes without saying that a story can hardly be represented in the form of Cartesian graph. Lévi-Strauss' typical use is rather that of textual tables, presenting what he calls "transformation groups", visualizing the links between stories that are in some way related, and often describing many differences as due to the effects of a primary transformation. We can therefore read the differences between mythical tales, for example belonging to two neighboring ethnic groups, not in terms of differences in those *two particular stories*, but as traces of the transformational relationship that binds the two mythological systems. If one myth seems somehow to deform another, the deformation can be attributed to the underlying semiotic fields. The "semiotic fields" that I am referring to are not, of course, abstract entities of logical nature but socio-cultural realities, that correspond to alternative ways of viewing the world, symbolic systems in conflict with one another, behind which we can see the presence of ideologically antagonistic power centers. As in the case of Thompson's animal species, we can also explain the form taken by the stories as dependant on the action of dynamic factors, almost the hint of a diagram of forces.

There is no need to emphasize that textual theory emerges greatly innovated, or that this perspective seems particularly pertinent precisely in that it no longer conceives of the cultural system as a sort of collection of statically and individually defined texts, but as a network of connected texts and textual portions. In this different perspective the generation of the texts is transversal, implemented by deformation and comparison, reaction and opposition. Since all textual content is related to relational reading, the difference, which in Propp's model (and partly in that of Greimas) separates the formal or syntactic level from the content or variable components level, disappears. The formal dimension, rethought in terms of a relational plan and an active process, *invests all levels and components in the same way*. Nothing escapes.

## 7.6 Narratives in Conflict

Unfortunately, Lévi-Strauss, despite having left us many suggestions and directions, did not leave us a systematic analysis to satisfactorily illustrate these concepts. This was also due to his method, characterized by a strong centrifugal development, which led him to explore an increasingly wider set of texts and cultures. Personally, I have conducted a test study on one of the specific cultures and specific delimited mythological complexes studied by Lévi-Strauss, and at this point I find it useful to make a very briefly reference to this research,[4] so as to not leave the reader lost in the generality of abstract concepts and models. The fact that we are addressing the Bororo myths, including the myth from which the grand construction of *Mythologiques* starts (the myths considered correspond to, in Lévi-Straussian numbering, $M_1$, $M_2$ and $M_5$),[5] is superficially interesting. The fact that in this case there is a conflict between different symbolic systems that oppose each other within the same community is more relevant. We think of this kind of phenomenon as belonging to complex societies, but it can also be present in societies

[4] The study is included in the second chapter of the third part of the Ferraro (2001) volume. In that research I resumed a part of Lévi-Strauss' analytic material, uniting it with ethnographic information from other sources.

[5] The analysis of these myths lies in the first chapters of Lévi-Strauss (1964).

we consider "simple". Indeed, it is significant that the same missionaries who had contact with this population were puzzled by the contradictions that they encountered in indigenous beliefs, particularly with regard to the explanation of disease causes (as well as that of the origin of storms, which we will discuss in a little bit). Antonio Colbacchini (1925) suggests that there was in fact an overlap between an original belief system and another set of substantially foreign beliefs (and, it should be pointed out, institutions), probably imported from another tribe. We can therefore assume that we are faced with two different systems of semiotic coordinates in which stories are inserted that together make up the narrative heritage of a defined population. Even more interesting, however, is another consideration. The two symbolic and narrative systems that come into clear conflict in the Bororo mythological universe, despite sharing much textual construction material, reveal a profoundly, constitutively different nature.

Let's center our attention on the question of causes of disease. Two myths in particular belong to what we recognize as the first of the two conflictual semiotic fields, one centered on the origin of watercourses, the other on the origin of diseases. These stories talk about mythical times characterized by a total absence of order, and later about the establishment of some kind of order, universal, cosmological, ethical and social at the same time. The era of the disorder is marked by incest, greedy people who do not share food with others, and by the absence of those waterways that organize the material and symbolic geography of the territory, and by the correlative absence of funeral rites, resulting in the reprehensible practice of keeping the bodies in pits dug under the huts. The creation of the waterways and the parallel establishment of burial rituals (in the water of the rivers, in fact) intervene to bring an order which is then supplemented by the establishment of a separate and corresponding kingdom of the dead. This kingdom will be entrusted to two village leaders belonging to one of the two *social halves* that compose the Bororo society, while the world of the living will be ruled by two village leaders belonging to the opposite *social half*. As you can see, the game of symbolic symmetries can skillfully cover what we would consider the givens of an asymmetry of power. As for the origin of the disease, this is presented as a result of antisocial behavior. A woman, after having caught a lot of fish, ate them immediately rather than sharing it with relatives. Her belly swells as a result, and when the woman returns to the village her body gives birth to every kind of disease.

These references bring us back to an obviously broader and very well organized system, with distinctions and symmetries, exchanges and reciprocal duties, lines of separation and regulated transition processes. Its hold on reality depends also on the ability to enclose the entire universe in a global model, including spiritual components. In fact the symmetrical relationship between world of the dead village leaders and world of the living village leaders is very significant, as well as the latter's task to hold dance rituals in which they symbolically embody the dead. Symmetrically, even here, deceased members are impersonated from members of the opposite social *half*. The rules of incest are just as much part of this overall conception of the ordering of the real (forbidden to have any proximity with someone of your own *social half*), as is the explanation of the disease, seen as a consequence of the breakdown of the mechanisms of mutual exchange that this worldview is based on.

To provide an idea of the syntagmatic arrangement of events, the *myth of Baitogogo* ($M_2$) describes how in ancient times there was a boy who had reached the age of initiation, but did not want to leave his mother. One day he secretly follows her in the forest, where he sees his incestuous betrayal with a man belonging to his own *social half.* The father kills his wife and rival and buries them in a pit dug under the hut. The child turns into a bird and inseminates a rapidly growing tree on his father's shoulder (notice this irruption of the vertical dimension, that appears hostile and a little out of place in this story). The man runs away from the village, and the tree is transformed, giving rise gradually to waterways. In this watery world, he founds the happy kingdom of the souls of the dead and creates the necessary ritual tools for funeral rites.

Other stories refer to a different perspective. What is known as *the myth of Geriguiguiatugo on the origin of storms* opens with a young man who, at the moment he was about to be initiated, follows his mother into the forest and rapes her. The boy's father (who is one of the two village leaders) seeks to cause the boy's death in various ways, but the boy escapes all trials. From the beginning, the stories seem to speak to each other from a distance, as if resuming and disproving the content of the other. For example, this is true for the symbolic element of water, which we saw occupy a fundamental place in the founding support of universal order in the stories of the first group. Our Geriguiguiatugo is instead immune to water, thanks to the support of birds or the ability to transform himself into a bird (just like the son of Baitogogo in the other story). But the ability to transform himself into a bird seems like a first reference to another social institution, that of the fearsome Bororo sorcerers, the *Bari.* And on this road we can come to understand that this story, in opposition to the previous ones, is supporting an alternative world view, in which the social order (symbolized by the rule of incest and placed on a substantially *horizontal* plane) is not only markedly diminished but particularly dissociated from the cosmological order and the *vertical* dimension, called on to connect the world of men to that of the spirits. In this second vision, for example, both storms and diseases are sent by spirits of dead sorcerers (who dominate the *world above*), as punishment for lack of respect, and therefore have no connection with issues of anti-social behavior. The end of this myth recounts, not by accident, how a storm puts out the village fires. Then the young Geriguiguiatugo, transformed into a deer, attacks his father and kills him. The story ends in this manner with the killing of the village headman, highest chief of the social order, by a stag, the animal in which it is believed that the dead sorcerers reincarnate. In addition, we find the establishment of the avenging power of the storms: two clear references to the sinister power of the *Bari.* The stance taken by this tale is therefore unquestionable.

The outline of each story, however, seems to take on real meaning only if it is seen as a polemic transformation of other stories. Even though we can not go into details, it should be emphasized that, if the myth of Geriguiguiatugo "is subjected to" the conceptual structure of the adversary's mythology and builds upon it, this is also true for the Baitogogo myth, that must create a place for a unique bird that seems to appear as a sort of citation of the opposing story. We can speculate, in effect, that the comprehension of these stories requires the listeners to mentally compare the corresponding textual structures. In a sort of battle for the ownership of the traditional mythological nucleus, each party tries to include the opponent's themes and narrative configurations, so that both the "chief's" and the "sorcerer's" mythologies seem to reproduce common patterns, distorted and reorganized in order to respond to the symbolic coordinates that define the two antagonistic *semiotic fields*.

In essence, the "chief's" mythology outlines the logic of a world founded on a pervasive *symmetry*, a symmetry that is actually easy for the natives to find in their experience, starting with the fact that the social universe is clearly regulated by the symmetry between the two *halves* of which it is composed. How this reciprocity extends also to the *metaphysical* relationship between the living and the dead is rendered palpable by the participation in ritual dances, in which each person plays a deceased person from the opposite *half*. On the contrary, the system of stories and beliefs that support the power of the sorcerers is characterized by a strong asymmetry and by the decisive, terrifying controlling power that the spirits exert over the natives, and in particular on their health. The preeminence of the *vertical* dimension (from which the destructive water of the storms comes) is highlighted, together with a *subjective and individual* dimension. It is a distressing world, marked by an intersection of premonitory and expiatory tensions.

The conflict between the two different codifications of the world is in our eyes even more interesting because it confronts two very different narrative construction logics. The first follows modes typical of a pre-literate mythology, based on a play of analogical correspondences between the different aspects and levels of reality. The personal stories have a very limited *perspective* dimension. What matters is the perfect logical hold of the *map*, that the events placed "in mythical time" serve only to establish, consolidate and stabilize. Errors and shortcomings directly correspond to that which could unravel the system: leaving ones appointed position; not respecting symmetries and reciprocity. The regular *systemic grid* is exactly what must be safeguarded. The other rationale is presented as instead based primarily on a rift that distinguishes this all-encompassing grid into two distinct areas: the horizontal order of the social universe, acknowledged but delimited; and a more decisive order, vertical and supernatural, which plays on *emotions and subjective plans*, where the principle of reciprocity turns into revenge, error into sin. The harmonious, static games of multiple analogical correspondences typical of the myth, concede the field to narrative settings closer to those that we are more familiar with.

The study of the reciprocal transformational relationship between the stories included in the two respective semiotic fields is therefore doubly interesting, because in addition to showing how the myths (somehow like Thompson's famous fish) change their form, passing from one frame of reference to another, it also helps us observe the transition between two different modes of narrative construction. In as much as we can see in these allusions, the first mode, decidedly situated on the *map*, bets everything on the strength of the order of its own coordinate system, trying to avoid events that will disturb its regularity. The second argues instead that the orders are numerous and hard to decipher, and that life is a *journey*, a set of actions that can be right or wrong, each of which involves consequences. The process, we would say, dominates the map. But the natives, to whom in all likelihood the relationship between the two systems in conflict is not so obvious, continue to repeat stories that seem to *objectively* resume, replicate content, vary episodes, as in a game of mirrors to which one can superficially assign a only formal value.

## 7.7 Aliens and Bohemians

Levi-Strauss' theory was not developed through the study of oral narrative texts by chance, where the continuous modification of the stories is of course facilitated by the absence of a written reference version. This does not mean, however, that it is not applicable to stories from our recent culture. In a recent research project (Ferraro and Brugo (2008), for example, I investigated the evolution (from the late nineteenth century to the beginning of this century) of stories and figures that were called on to symbolize *absolute evil*, such as vampires, monsters or aliens. In fact, it was possible to identify a coherent and potentially global variation logic, exactly as if the overall deformation of the cultural coordinate system had determined the parallel transformation of the stories, even in cases where the authors were sincerely convinced of *repeating them without changing anything*. This alteration, to give you an idea, is related primarily to the development that starts from the collocation of evil in an impersonal and external place, leading to the belief that we ourselves create our own monsters, indicating a relationship that is no longer separate but involved.

For example, if we consider the four films in the *Alien* series, directed by four different and important directors from 1979 to 1997, we find that the same story was basically told four times, changing gradually in relation to the cultural system in which it was inscribed. This phenomenon is conceptually very close to the most crucial aspects of the theory of transformations presented by Thompson in *Growth and Form*. Even though the four episodes of the saga maintain a consistent, common ideological position—a basic anti-establishment, anti-religious, and in its own way anti-capitalist attitude, hostile to any combination of scientific and commercial interests—an evolution occurs, which in some respects leads to a true reversal of the original symbolic structures. The relationships between these texts could indeed fit very well into transformational schemes used by Lévi-Strauss. This could include the figure of the alien. In the first film, the monster penetrates the human being in a sort of gruesome sexual act, reducing it from subject to a mere *object*, a mere instrument for its reproduction. In the end, changing gender and role, yet with an absolutely compelling transformative logic, the alien appears as the daughter generated by the protagonist. And this one, the protagonist of the story, is defined from the start as an *object*, who only in retrospect can access *human* status.

If we look carefully, we notice that it all depends on a shift of the plane of meaning. The placement of the 'evil' that we addressed earlier essentially changes, and with it the conception of causal agency. There are no passive victims, but entities that are architects of their own condition. Once this is understood, it becomes clear that all of the transformations are clearly systemic, starting with the one that, while maintaining the reproduction of aliens through human beings as a fixed process, shifts the focus of attention from fertilization (passive, for the human being) to generation (active). There is a polarity change from *male to female*, which also corresponds, on the negative side, to the name change of the oppressive controller computer system from Mother to Father. Following this logic, we find a shift from the natural to the artificially built, from an identity given at birth to an identity actively acquired, from a negative to a positive perception of hybrid human/non-human entities. We shift from fear of the alien to fear of the human, from a simple and integral dread to an anxiety mixed with compassion and tenderness. We abandon the idea of a return to a "home", understood as the comfortable world that we came from, for the arrival in a world that seems both our "own" and unknown, projected into a still unreadable future… We find ourselves in front of the transformation tables that fill the pages of Lévi-Strauss' research on American mythologies. The narrated story repeats an almost obsessive formula, rewarding the viewer with the pleasure of repeatedly experiencing the same adventure. Yet the meaning of the story undergoes a sharp twist, as a result of changes in the containing framework, due to a sort of contraction that folds the *external* pole back on the *internal* one. This story, which had an important place in the collective imagination of the time, in short, gives us a clear example of the way in which textual forms undergo changes that can only be understood if seen as effects of a single comprehensive transformation, giving rise to a whole set of related changes. Moreover, this comprehensive transformation is in itself well explained as dependant on the action of defined cultural dynamics. What was difficult to grasp through the use of traditional generative semiotic methods, seems rather clear in the light of a theoretical model with a transformational and systemic approach.

I would like to close these reflections, however, by citing a case that is even more particular and intriguing, for our theoretical speculations.[6] At the decisive moment of the transition from a somewhat tired nineteenth century to the triumphant modernity of the twentieth century (in 1896), an artist who was curious about everything new, open to international culture but at the same time undoubtedly linked to the romantic tradition, Giacomo Puccini, created *La Bohème*, which appears as both one thing and its opposite. The story is old and well known, coming from a successful comedy 50 years earlier. It speaks of a world, people and times long past, while resuming a model that is the epitome of a certain kind of stories. In fact it draws inspiration from the inexorably moving archetype, the literary *femme fragile*: pale and sickly, linked to cold and the moon, related to flowers and obviously ephemeral, Mimì is destined to unrelenting agony. The story, thanks to its setting in a bohemian attic where four young artists live, has all the necessary requirements to represent the triumph of romanticism. And yet, when it was

[6] Reiterating briefly the findings of a larger study, cfr. Ferraro (2009).

presented, many perceived it as exactly the opposite: an astonishing break with romantic opera traditions, an unprecedented concession to the new revolutionary style of realism at all costs. The judgments of the time are striking: for some the story is cloying like the music that accompanies it; for others it is a miracle of fine complexity; for some it is a backward and trite melodrama; for others still it is surprisingly sophisticated and innovative. It cannot even be said that the judgments of a certain kind can be attributed to conservative critics and others to liberals, because the opinions are mixed in a quite symptomatic confusion.

Puccini developed the story with the two authors of the text, who, not surprisingly, were also twofold in nature. One, Giuseppe Giacosa, was a classic and elegant man of letters, while Luigi Illica was an irregular and innovative poet. The story was created through endless bickering and negotiation, cut through with ellipses, narrowed to four flashes isolated in time. Therefore the story ends up perfectly ambiguous, and the figure of the protagonist is permanently unreadable. On the one hand, the work stands as a text that is properly inscribed in nineteenth-century culture, but on the other, it is also readable as a venomous parody that takes distance from nineteenth-century culture, fully set in twentieth-century models. In this case, we can truly say that the text exists essentially as a remake, as a citation of a way of feeling and as a transformation of other texts, which at the same time re-plays and rejects, perpetually suspended between moving sentiments and cynicism. Do the two protagonists really live a touching love story, cut short by an incurable disease? Is Rodolfo really a talented undiscovered writer, taken by his own poetic inspiration? Or are they instead two seedy characters? Could it not be that Rodolfo is a mediocre, unreliable, and distracted man; unable to care for the woman he claims to love, aware of his own literary failure, parked in an attic while he awaits an inheritance from his rich sick uncle? Couldn't Mimì be a frivolous girl, ready to crawl into bed with an old and wealthy man, in order to leave the miserable attic to which she returns only when she is already dying? The two truths, with the two opposing identities of the protagonists, are both present. The spirit of the past and the spirit of the future share the same spaces and the same gestures, even the music is both moving and anti-naturalistic. The viewer can perceive everything as terribly trite, or as surprisingly new and alienating. Puccini created, in essence, a kind of *meta-text*, built on "romantic" stories and ways of feeling. Properly, neither the old world of bohemian attics nor the new world that should replace it is represented. More precisely, the *passage* from one to the other is staged, seizing the moment in which the one is still present and the other is not yet fully established. What in short is shown is the change as such, the jump that leads from a cultural system to another, as well as the problem of mutual illegibility that this entails.

It is evident, I think, the pertinence of this example to our discussion. The fact is that the transformational relationship between one fish and another does not pose cultural problems, but things are very different when we talk about the passage of stories from one semio-cultural system to another. *Bohème* is, of course, a particularly fascinating case, but far from extraordinary. Indeed, the most interesting aspect of using Thompson's model in a narratological context, is perhaps that of focusing attention on those texts that, rather than definitely belong to one or another system, pose themselves *crosswise* so to speak, simultaneously settled in multiple reference systems. They present themselves, of course, as complete and defined entities and are therefore perceived *as individual objects*, but in fact possess multiple identities, dependent on the different reference systems in which they are inscribed. The use of Thompson's model allows us to imagine fascinating extensions of semiotic methodology, helping us intuit how the study of relationships between systems could be more effective and valuable than an analysis of objects conceived as such. We can pay the price of giving up objective and immanent definitions of textual constructs, in order to better understand the dynamics of change and the interaction of the systems of semiotic coordinates that ultimately define their meaning.

## References

Almén, B.: A Theory of Musical Narrative. Indiana University Press, Bloomington (2008)

Colbacchini, A.: I Bororos Orientali "Orarimugudoge" del Matto Grosso (Brasile). S.E.I, Torino (1925)

Ferraro, G.: Il Linguaggio del Mito. Valori Simbolici e Realtà Sociale Nelle Mitologie Primitive. Meltemi, Roma (2001)

Ferraro, G.: Al buio non si trova. La Bohème alla luce della sociosemiotica. In: Pozzato, M.P., Spaziante e L. (eds.) Parole Nell'aria. Sincretismi fra Musica e Altri Linguaggi, pp. 229–253. ETS, Pisa (2009)

Ferraro, G: Attanti: una teoria in evoluzione. In Lorusso, M., Paolucci, C., Violi e P. (eds.) Narratività. Problemi, Analisi, Prospettive, pp. 43–60. Bononia University Press, Bologna (2012)

Ferraro, G.: Fondamenti di Teoria Sociosemiotica. La Visione "Neoclassica". Aracne, Roma (2013)

Ferraro, G., Brugo, I.: Comunque Umani. Roma, Meltemi (2008)

Lévi-Strauss, C.: Le Cru et le Cuit. Plon, Paris (1964)

Prince, G.: Narratology. The Form and Functioning of Narrative. De Gruyter, Berlin (1982)

Propp, V.: Morfologija Skazki. Academia, Leningrad (1928)

Propp, V.J.: Istoriceskie korni volšebnoj skazki, Leningrad (1946)

Thompson, D.W.: On Growth and Form, II ed., vol. 2. Cambridge University Press, Cambridge (1952)

White, H.: The Content of the Form. Narrative Discourse and Historical Representation. Johns Hopkins, Baltimore (1987) [illegible] expose themselves, so to speak, simultaneously sorted in multiple reference systems. They present themselves, of course, as complete and defined [illegible] the different reference systems in which they [illegible] Thompson's [illegible] [illegible]

## References

[illegible]

# Part III
# Immanence in Semiotics

# Chapter 8
# Semiotic Immanence and Formal Options

**Francesco Marsciani**

**Abstract** At the beginning of Prolegomena, Louis Hjelmslev arrives at his own version of linguistic theory in a brief journey, eliminating in succession a series of explanatory criteria on the nature and function of language as a phenomenon shared by humans, and as a signifying practice.

## 8.1 Immanence and Form in Hjelmslev's Linguistics

At the beginning of *Prolegomena*, Louis Hjelmslev (Hjelmslev 1943) arrives at his own version of linguistic theory in a brief journey, eliminating in succession a series of explanatory criteria on the nature and function of language as a phenomenon shared by humans, and as a signifying practice. The reason behind this short journey is the need to purify the theory of any criterion or rational that transcends the linguistic phenomenon in itself, as well as the intention to clarify the levels of relevance regarding the set of categories and concepts to be used in the construction of a "science of language" worthy of the name. The same Hjelmslev considers these levels of relevance as planes of immanence.

The construction of a "scientific" theory (in Hjelmslev's sense) is rendered possible by starting from a thus recognized plane of immanence, through the development of a system of concepts, based on structural functions, whose main characteristic is that of responding to a strong criterion of inter-definition. The plane of immanence on which the concepts of the theory are organized, thought up essentially according to the structural form of an "internally consistent" relational system, for this very fact then assumes the traits of an autonomous formation. It is in many ways independent from the appearance of linguistic phenomena with which the linguist, the "scientist of languages", is obliged to work with. It is organized following an orientation that takes a clear position in the definite

F. Marsciani (✉)
Università di Bologna, Bologna, Italy
e-mail: francesco.marsciani@unibo.it

 
A. Sarti et al. (eds.), *Morphogenesis and Individuation*,
Lecture Notes in Morphogenesis, DOI 10.1007/978-3-319-05101-7_8

epistemological alternative between inductivism and deductivism. The plane of immanence that Hjelmslev outlines is a plane in which the deductivist principle is the criterion for conceptual and systematic construction, according to which the specifications proceed from the general axioms, and are articulated according to a progressive “development of the possible”.

The scheme is quite simple and is in substance what following semiotics, structural and generative, has essentially preserved. Such a scheme, in any case, is realized on the basis of a radical displacement of the relationship between descriptive language (metalanguage or which ever formal model one intends) with the language object on which the described phenomena stand. To be precise, “phenomena to describe” are no longer even exactly the object on which to apply a theory that respects the structural nature of a plane of immanence such as the one we just described. We are referring instead to the construction of possibilities, or rather the constitution of phenomena in their essential nature, namely the production of spaces likely to be filled, occupied or realized by a completely variable and undefined number of manifestation substances. The Hjelmslevian option is very clear and persuasive in this regard. Hjelmslevian immanence prevents the risk of an “internalist” version, so to speak. This is particularly interesting: in order to arrive at the plane of immanence, Hjelmslev takes a route which requires the elimination of that which, as we mentioned earlier, seems to have to present itself with a transcendent nature with respect to the phenomenon of language (those which the same linguist identified as historical, sociological, psychological, economic, etc., criteria), namely extrinsic elements. The plane that he reaches is not only the place where the opposing elements lie, that is, the intrinsic elements (elements authentically linguistic), but the one in which the theory finds, more radically, a transcendental horizon. It is a entirely immanent transcendental, of course, and in the same way it is also a field populated by the elements intrinsic to the linguistic fact. The type of object that is the linguistic fact as a phenomenon, in the sense that it continues to apply the general rule that determines the strictly linguistic nature of a language phenomenon. The criterion that makes us ask, first of all, what makes a phenomenon of language a phenomenon of language, what are the characteristics that make it, allow it or force it to be, simply what it is. In any case, the transition presents itself at the same time as a reversal of the problem, a reversal of the relationship between model and object. The transcendental option that defines the immanence of Hjelmslevian linguistics, particularly the immanence belonging to structural semiotics from which it derives, requires that the delimitation of the ontological field of relevance of the “science of language”, its regional ontology (according to phenomenological terminology), does not depend on a horizontal selection, so to speak, of the boundaries and distinctions of the relevant classes, but depends on a production of objects, a constitution (again according to phenomenological terminology), which defines the possible on the basis of its conditions of possibility. In this manner, we find a very technical and operational notion of transcendental applied, namely a notion that refers to anything that acts as a “condition of possibility”, in fact, sufficient conditions of existence. Thus the phenomenon of language, if properly constituted as an object of an immanentist linguistic science, responds to criteria that define it on the basis of functional definitions of the conditions of possibility. A classic example: we can not speak of the phenomenon of language if we cannot find the two planes "inside it" that, a priori, define any phenomenon of language, the classic two planes called expression and content; as exemplary and in the same manner, we cannot speak of the phenomenon of language if we cannot find "inside" each of the two planes, the two axes that a priori define the functioning of meaningful articulations, the two classical axes called system and process. In this manner it is the theory, formed by the possible conditions of a given type of phenomenon, which establishes a priori the nature of that type of phenomenon as appropriate for the objectivity of which theory itself provides the description.

In developing this scientific option -which is centered on the development of a plane of immanence whose traits are, we have seen, the search for intrinsic elements that can define the phenomenon in question as a suitable object, and at the same time is elaborated in terms of a set of transcendental conditions of possibility- what is produced in net and undeniable terms is a gap or a radical departure from the empirical manifestation plane and the set of concepts that account for its relevant organization. Hjelmslevian immanence is understood as the extreme opposite of manifestation. It is no longer the opposition between immanence and transcendence that matters, but rather an opposition between what appears, the empirical experience of the use of language, and the set of the transcendental conditions of possibility of that same experience. This, that we have called a gap that is produced between two ways a phenomenon of language is significant, corresponds to that which Hjelmslev considers a unilateral presumption between substance and form. Language is a formal hierarchy of functions that theory deductively reconstructs, and the substances that make up the empirical appearance of the phenomenon (in the case of verbal language, for example, the association that exists between a particular sound, or acoustic image, and a certain meaning, or mental image) are secondary and derivative in as much as they are "realizations of possibilities". The substance, in this sense, unilaterally assumes the form because it is the form that guarantees the nature of linguistic phenomenon, that provides the "linguistic sense", the value of language.

It is well known that the whole "scientific vocation" of linguistics and structural semiotics is based on this option, an option on which Hjelmslev insists with conviction and which has had a considerable fortune in the subsequent history of the linguistic and semiotic disciplines. The formal aspect of the theory is heavily marked, and the a priori traits of a deductivist and transcendental conception of the theory itself are justified and further reinforced in this manner (it is on this basis, for example, that it becomes possible to set up analysis procedures in according to the strict criteria of the "empirical principle": consistency, completeness and simplicity). The plane of immanence is a formal level, it is the place in which forms are articulated, i.e., relations and functions. What happens in the world, so to speak, could not make sense, could not be a linguistic or semiotic phenomenon, if there was not a formation of possibility capable of organizing facts as concrete cases of formal possibilities.

From here we recall a number of Hjelmslevian assumptions (practically adages): there is no text without language; there is no semiotic without distinction between the two planes expression and content; there is no element, entity or component identifiable without a formal network of organized relations.

## 8.2 Immanence and Form in the Structural and Generative Semiotics of Greimas

The formal options of Hjelmslevian linguistics have kept all of their value in semiotics. Despite the growth of strongly critical perspectives regarding the so-called formalism, that in recent decades have produced a reversal of perspective in the wake of criticisms of classical structuralism and growth of a cognitive perspective, structural and generative semiotics certainly remains anchored in the dominant spirit of the Hjelmslevian immanentism. Simply, but also radically: if there is not a formal system that realizes it, the appearance of the phenomena of meaning (to be understood as signs, texts, speeches, uses, messages, selections, percepts and as concepts) can not in any way emerge compared to a confused and incomprehensible becoming; the intelligibility of the same phenomena, its own identifiability, can not be separated from its previous creation, from a relational structure that assigns defined places, from a systemic topology that establishes identity value. The formalizing option that semiotic accepts and develops, however, can not in turn separate itself from a clarification of that which in Hjelmslevian linguistics is recognized as a sort of difficulty: can the immanent nature of the theory and its characterization in a substantially transcendentalist sense converge without any friction? Can they simply coincide without the need to be revisited to some extent in their more traditional meaning? Better yet, if we reflect carefully on what conditions may permit the two meanings, the two immanent and transcendental aspects -aspects of the same essential ground on which the whole theoretical system rests- to coexist and coherently overlap, then perhaps we need to articulate the two moments between them in such a way to not allow the formal aspect to harden into a logistic scaffold with claims of ontological deduction, i.e. the actual production of the facts of language (as often happens in normative grammars) and to the same time does not permit the transcendental aspect, in its part, to take *raison d'être* from a priori categories that are also "realistic", constituting an alleged conditional nature of some entity located at the beginning (human spirit, intellect, sensitivity, and whatever else philosophy may have mentioned in its fruitful history of ideal productions). In order to articulate the immanent aspect of the conditions of possibility together with the transcendental aspect of the immanent forms we need to define the grounds for validity that could not be based elsewhere, not at any depth or at any height, not in any destiny, nor in any origin. The immanent criterion would ask the transcendental to not transcend phenomena, and the transcendental criterion would force immanence to take form and not submit to the substantive adventures of the phenomenon to which it belongs. This could be the great enterprise of an informed semiotics, a semiotics capable of returning precisely the nature of its vocation to the same scientific vocation, its nature as a way of becoming, of continuous incremental transformation of reasons that arise from the same place where they plan to, or have to, go back to account for their ever renewed forming of life. Before a similar undertaking could take place, or even be planned, the relationship between immanence and the transcendental must involve the two opposite terms of form and substance in mutual dynamics. A risk that formalism has actually taken, at least within the scope of the theories of language, was to think about the opposition between form and substance more like an informal separation than a legal distinction, or more like an opposition between two entities of different nature rather than as two functives of an essential function. One would need therefore to adopt a less dualistic conception, so to speak, more phenomenological—to imagine both the object to be known and the science that examines it, with its constituent categories, as both floating in the world-of-life—and open a theoretical space in which the relationship between the two poles of the relationship can be incessantly translated, regenerated from time to time, respecting the fundamental value that is the structural principle of the primacy of the relationship on terms. This means that form and substance must remain the functives of a relationship that ensures that the substance, any substance, is what it is thanks to the form that organizes it and decides its relevance, but that their relationship does not connect two entities made in advance, two independent qualities—against any structural reason—rather two moments of a ceaseless transformation, of a "put in relation", of a mutual determination. If this space is needed, then, its construction coincides with the theory itself, i.e., the set of constitutive categories that are organized in structured system. Here the structure is above all the organization of the conditions of the possibility of the phenomena, but it immediately appears, exactly at the same time, that the form of the scientific description of the same things, the form controlled, by inter-definition, by the necessary practice and for that universal that consists in paraphrasing, objecting, transforming the given meaning in new meaning, in new signification. In this space that is the theoretical space, a curious or perhaps admirable coincidence occurs, a not trivial coincidence for the determination of the scientific vocation in general: the form of the theory and the transformation of the given meaning in new significance do not belong to different moments of the constitution, but rather are distinct moments—based on the "vocation"—of the same interpretive work, of the same practice of knowledge or appreciation or, if you will, seizing, on the world of phenomena invested with the value of sense. A semiotics thus conceived becomes the ground on which to practice the science as a "vocation", in fact, i.e. as activities of forming given substances, of formal organization of the empirical given, of predisposition of intelligibility criteria, without which, as we have already said, there is no meaning.

Greimas showed caution regarding any type of metaphysics taken as a point of eventual reference for a decision on the nature of the immanent plane of semiotics, and preferred to adopt a "low profile" operational criterion, so to speak. It may be interesting to re-read a brief quotation from the "Immanence" entry of *Sémiotique*:

> L'affirmation de l'immanence des structures sémiotiques soulève alors un problème d'ordre ontologique, relatif à leur mode d'existence : tout comme autrefois on s'était interrogé, à propos de la dialectique, pour savoir si elle était inscrite 'dans les choses' ou 'dans les esprits', la connaissance des structures sémiotique peut être considérée soit comme une description, c'est-à-dire comme une simple explicitation des formes immanentes, soit comme une construction, si le monde est seulement structurable, c'est-à-dire susceptible d'être 'informé' par l'esprit humain. Il nous semble opportun, pour écarter de la théorie sémiotique toute querelle métaphisique, de se contenter de la mise en place de certains concepts opératoires [...] (Greimas & Courtés 1979)

The alternative thus placed between the two options is clear, and it is, moreover, an effectively metaphysic alternative, which is very reminiscent of the classic alternative with regard to the nature of the structure as key concept for a whole epistemology of which we know the fortunes (the structure is in things or in the description that we give them?). Now, the caution towards any type of metaphysics may be treated with reservation through the argument that, among the indefinable with which a theory is equipped, necessarily taking them as abstract, preliminary and general axioms from which to derive the deductive chains, it would be difficult to protect from any type of metaphysics (even the rejection of metaphysics, as they say, is itself a form of metaphysics). What begins to appear of greater interest, however, is the possibility that a consistent and mature semiotic theory could take a step forward regarding this same net alternative, and consider its field of operation, its postulates as well as its methodological and descriptive derivatives, inside a dynamic of ceaseless formalization of substances, and think in this manner the formality of which it is equipped, rather than as the creation of formal precepts justified elsewhere (necessarily in transcendence, relying mostly to the formalization of logic) as the result of a theoretical control exerted on the chains of the substantial transformations. In other words, it should not be a decisive difficulty to think of this type of semiotics as a immanent "shaping" of the transformations of meaning.

It's true, in fact, that Greimas clearly came upon this type of solution ten years before the publication of the previous citation. It is sufficient to look at the introductory essay in *Du Sens* of 1970, "Sur le sens" (Greimas 1970), to find precisely this inspiration and to appreciate, in particular, the conversion between the formality of metalanguage and immanent form of transformations. So, it is on this same path that one can imagine the immanent transcendental in semiotics; not yet a transcendental that draw its own conditions of possibility from a given structure or repertoire, nor an immanence that thinks that the forms that ensure the significance of phenomena are hidden in the substance of the event, but rather as a formation that determines the transcendental structures (of the forms) that are needed for their unpacking and at the same time a categorical apparatus that can be discerned in the "real" chain of transformations from substance to substance. Since some kind of realization of substance is not feasible that is not informed in some measure, enhanced by a form that embraces the very aspects that make it what it is, Greimas is correct in raising the issue of what nature this form has; and yet this form may well be understood as that which the transformations of substances lets go, as constant aspects of the variants and as sediments of continuity, so to speak, that progressively pave the way for a particular field. We are referring to the field of comparability between different objects, that field that each time—and each time it is renewed—is constituted as the plane of relevance. The formal conditions of possibilities, therefore, are immanent to the substantial projects of the values at stake, and these same transformations decide, in their presentation as events, such forms must be filtered like drip coffee, if you will forgive me the expression.

There is no semiotic immanence if not within this formative range (forms, in fact, are resolved in a formativity that is always active because they are intrinsic to trans-formations). Into this same range the theory attempts a coherent inter-definition, systematic as far as possible, attempting to maintain a guiding "scientific vocation", and a dynamic form of rationality, which guarantees the exchange and communicability of the results.

All of the so-called Generative Path is precisely an attempt to inter-define the possibility conditions of meaning, stretched as it is between an instance of the generality of conditions, which are open to possibilities, and an instance of adequate and acute description of the live transformations that ceaselessly create form.

## References

Greimas, A.J.: Du sens, Paris: Seuil (Eng. Trans. On Meaning: Selected Writings in Semiotic Theory. University of Minnesota Press, Minneapolis, 1987) (1970)

Greimas, A.J., Courtés, J.: Sémiotique: dictionnaire raisonné de la théorie du langage, Paris: Seuil, (Eng. trans. Semiotics and language: an analytical dictionary. Indiana University Press, Indiana, 1983) (1979)

Hjelmslev, L.: Omkring Sprogteoriens Grundlasggelse. Munskgaard, København (1943), (Engl. Trans. Whitfield, F.J. (ed.): Prolegomena to a Theory of Language, University of Winsconsin Press, Madison (Wis.), 1961)

# Chapter 9
# Structural Syntax and Quantum Computation: A Simondonian Approach

**Francesco Galofaro**

**Abstract** The general purpose of the article is to propose a dynamic model of the constitution of the plane of immanence, in which meaning is represented as an individuated form (*Gestalt*) through a process of formation (*Bildung*) starting from a pre-individual, undetermined semantic universe.

## 9.1 General Overview

The general purpose of the article is to propose a dynamic model of the constitution of the plane of immanence, in which meaning is represented as an individuated form *(Gestalt)* through a process of formation *(Bildung)* starting from a pre-individual, undetermined semantic universe. This process is characterized by an addressing function (Addresser as intended by Greimas), representing the introduction and the circulation of semantic values in the semantic universe. The model refers to various features of Gilbert Simondon's epistemology: individuation; meta-stability; and indetermination.

In order to explain the model's features from an epistemological position, in the first part of the article we will first sketch the history of the relationships between Morphogenesis, Structuralism, and Semiotics. This will introduce the opposition Gestalt/Bildung as a relevant category for understanding the dynamics of the formation of meaning.

In the second part of the article, we will see how, on the basis of linguistic data, structuralism contests the leibnizian notion of identity, establishing a parallel with quantum logics, and introducing the problem of indetermination, consistent with Gilbert Simondon's position.

F. Galofaro (✉)
Politecnico, Milan, Bologna, Italy
e-mail: semioatp@gmail.com

F. Galofaro
CUBE, Università di Bologna, Bologna, Italy



A. Sarti et al. (eds.), *Morphogenesis and Individuation*,
Lecture Notes in Morphogenesis, DOI 10.1007/978-3-319-05101-7_9

The third part resumes Greimas' actantial model describing the circulation of value in an immanent semantic universe. In as much as it is inspired by a computational generative approach, Greimas' formalism describe meaning as a *Gestalt*, which we can consider the static result of a metamorphosis between two meta-stable states. In particular, Greimas describes an addressing function, which introduces value into the semantic universe, and regulates its circulation.

In order to avoid the non-relational, static features of Greimas' formalism, and in order to describe the metamorphosis in itself with a formal model, the fourth part introduces Renè Thom's archetypal morphology, which is in fact more dynamic. Nevertheless, it opens different problems: first of all it is a deterministic model; second, it does not explain what happens in the case of the introduction of new values in the semantic universe; third, by explaining the morphogenetic processes through the manipulation of some control parameters, it simply shifts the problem about the origin and the nature of these parameters. He calls *saillance* ("salience") the value of an object; at the same time, this value presupposes something which can "intercept" it; Thom calls this instance *prégnance* (a therm which has been translated with the neologism "pregnance", or with "pregnancy").

For these reasons, the fifth part proposes a new model for the addressing function, which explains the realization of actantial structures using the tools provided by quantum computation (another simondonian idea). The introduction of a value into the semantic universe is seen as a wave function (on the *Bildung* side). At this level, the relational epistemology of structural linguistics and semiotics can be represented as an entanglement between the involved actors. The virtual presence or absence of a given value in the semantic universe is therefore represented in a formal model as a coherent superimposition of contradictory states; the realization of the value is represented through the wave function collapse. If the result is positive, then we obtain the realized actantial structures that are the geometrical boundaries of the value circulation (on the *Gestalt* side).

In the sixth part we discuss the implication of the proposed model as far as the constitution of the plane of immanence is concerned, from linguistic structures to any given experience of the *Lebenswelt.*

## 9.2 Semiotics, Structuralism, and Morphogenesis

Though René Thom's work undoubtedly represents the major intellectual exchange between semiotics and morphogenesis, the relationships between the two approaches to form are undoubtedly older, and do not always share mutual understanding. If the work of Goethe and D'Arcy Thompson played a strong influence on the structuralist weltanschauung, Propp's work, inspired by Goethe, instead met Claude Lévi-Strauss' criticism. As Thom is concerned, some features of his models—naturalization, determinism—do not seem in line with the more innovative features of structuralist philosophy.

### 9.2.1 Goethe's Metamorphosis

In order to overcome the ancient debate on preformism and epigenesis, which characterized the epistemology of his time, Goethe chose to dedicate his efforts to a theory of form, described as a phenomenic process not based on conjectural categories and principles. At the same time, his observations were based on a comparative method, which led him to investigate the deep structural unity of the living being. Nevertheless his idea of structure is a dynamic process, based on periodical transformations in the geometrical structure (*polarity*): his interests were directed toward the plasticity of the form. Leaf is not a static schema (*Gestalt*), but a dynamic development (*Bildung*)—cf. Mondella (1975:238). Goethe had no instruments with which to explain the epistemological framework of his observations, being incapable of solving the antinomy between idealism and empiricism. In a famous dialogue with Schiller, the poet invited him to consider his *ur-plant* as an "idea", a philosophical point of view that was in fashion at the time. Goethe's beliefs were focused on concreteness, and this is the reason for his suspicion of mathematics. Nevertheless, he developed exact methods that would have an impact on different fields, starting with linguistics.

### 9.2.2 Von Humboldt

Wilhem Von Humboldt was a scholar and a close friend of Goethe's. According to Cassirer (1945), he was influenced by the concept of polarity, and transferred his theory of organic types to the study of languages. The fact that Kant's research had neglected language left German intellectuals and philosophers such as Herder unsatisfied. Von Humboldt applied himself to languages as a concrete realization of the human spirit, studying them not from the point of view of their material features, but as the historical diversification of a single form. At the same time, he thought the origin of the difference between languages is related to the different way in which a culture structures its knowledge. Here scientific linguistics meets anthropology—cf. Geymonat (1975, p. 24). According to Cassirer, Von Humboldt's point of view had a strong influence on Viggo Brøndal's structuralist program. But this is not the only influence that Von Humboldt had on structuralist perspectives. As De Mauro (2001, p. 354) demonstrates, Von Humboldt had a great influence on Saussure too, with reference to the notion of *innere Sprachform* (inner speech form) and language as an *innerlich zusammenhängenden Organismus,* an internally coherent organism because of its interconnection. Cassirer's work contributed to the active diffusion of Goethe's among structuralist scholars. For example, Hjelmslev was very interested in the most up-to-date philosophy of science available at the time: Carnap, Husserl, and Cassirer also—cf. Caputo (2010, p. 36). The quoted article was originally published in *Word*, the journal of the Linguistic Circle of New York, directed by André Martinet.

### 9.2.3 Vladimir Propp

Propp headed a number of chapters in his *Morphology of the Folktale* (1928) with quotes from Goethe's work, which were omitted in the English edition. In debate with Lévi-Strauss—in Propp (1984)—the soviet scholar criticizes the choice of the editor, because these quotations have a great explanatory power on his conception of form. If the French ethnologist depicts the narrative system as a combination of elements, wherein, sooner or later, every concatenation will be attempted, Propp's conception is instead sensitive to the dynamic development of the structures in the diachronic dimension, which can be useful, for example, for dating different variants and explain them in relationship to cultural change. In my opinion, here we find again the distinction between the two Goethian notions of form (*Gestalt Vs. Bildung*). In fact, among the quotations we find: "the study of forms is the study of transformations" (*Gestaltenlehre ist Verwandlungslehre*). In the quoted texts, Goethe considers his ur-type as visible in the median section of every organic shape. As they belong to every living being, the laws of the form could even be used to invent new forms. These conceptions did not pass unnoticed: Greimas remarks how Propp uses the term "morphology" in a botanical sense. He also establishes a link between Propp's morphology and his actantial model—cf. "Morphology" in Greimas and Courtés (1979). Because of the conflict between Propp and Lévi-Strauss, this particular morphogenetic root is not emphasized in Greimas' work. Nevertheless his canonical narrative path, the summa of structuralist narratology, can be considered as a generalization of Propp's, even if from a critical position—cf. "Transformation" in Greimas and Courtés (1979).

### 9.2.4 D'Arcy Thompson

Even if Lévi-Strauss and Propp diverged on many points, it is important to underline how, in the work of the French ethnographer, Goethe and D'Arcy Thompson were as important a source of inspiration as linguistics—cf. Lévi-Strauss and Éribon (1988). According to Lévi-Strauss, the notion of "structure" can not be understood without the complementary notion of transformation. I'd like to highlight another interesting point, which Lévi-Strauss takes from Thompson's work: the *principle of discontinuity.* The potentially infinite number of transformations generates discrete differences, which do not only correspond to the exigency of taxonomies, but are also a feature of the objects we are classifying. Thompson compares it to *quantum theory.*

### 9.2.5 From Structuralism to Semiotics

As Guido Ferraro (2010) notices, the notion of geometrical transformation may have seemed distant from the linguistic notion of structure at the time in which

Lévi-Strauss used it. Nowadays it has become part of the topologic approach to meaning. Semiotics inherited many morphogenetic views and tools from Structuralism. According to Maurizio Del Ninno (2010), the consequences of the presence of Thompson's and Goethe's concepts in Lévi-Strauss thought will still produce fruit in the development of ethnosemiotic theory. We find traces of these roots in Greimas' actantial structure, and it is not an accident that René Thom considered them *archetypal morphologies*—see also Petitot, "Morphologies archétypes", in Greimas and Courtés (1986). Apart from that, Semiotics might be interested in the development of different morphogenetic concepts, such as the dynamic position implied by the opposition between *Gestalt* and *Bildung*, and the parallel between the discontinuity principle and quantum theory. As we will see in the next section, these features inspire this closer look to Simondon's work.

## 9.3 Notes on Individuation

Many of the epistemological ideas proposed by Simondon (2005) circulate in Semiotics thanks to Deleuze and René Thom. The reason why post-structrualism and Semiotics turned their attention toward Simondon's work are understandable in the light of the *discontinuity principle* and the related debate on differential structures.

### *9.3.1 Saussure and Difference*

At the base of Saussure's theory regarding meaning, we find only pure negative differences, without positive terms. Pure differences produce meaning. It is a well-known position:

> In the language (that is, a language state) there are only differences. Difference implies to our mind two positive terms between which the difference is established. But the paradox is that: in the language, there are only differences, without positive terms—Saussure (1993).

Saussure provide also an example of a pure difference without a positive term: the genitive plural of the Czech *zhena* (Woman), is *zhen:* the root in its pure form without declension affixes. Saussure was dazzled by the discovery of this feature in Czech declension:

> I am tempted to say that this word alone tells us far more than all that linguists have written on language, and philosophers on the fundamental workings of the sign-idea relationship—Saussure (2002).

Saussure's excitement seems strange: after all this declension is typical of many slavic languages in regards to the feminine and neuter grammatical gender, e.g.

Polish. The reason is related to the variety of thing we can consider as a "linguistic element":

> It is clear that in the language one sign is as good as another. Here there is none. *Zhena*, *zhen* functions as well as *zhena*, gen. pl. *zhenu* which existed previously. This example shows that only the difference between signs is operative. *Zhenu* works because it is different from *zhena*. *Zhen* works because it is different from *zhena*—Saussure (1993).

Thanks to the differential and positional nature of the system, we have a "nothing", which is "something" and carries information (in particular: gender, number, case). Today, Saussure's discovering of a zero-element may seem trivial to students, but at that time it was not. This feature of structuralist epistemology conflicts with the previous two thousand and five hundred years of western philosophy starting from Parmenide's ontology. In a system, even "nothing" can be an element, because in order to be an "element" it is not necessary to show positive qualities. It is sufficient to be different from something, and "nothing" is different from everything else. Saussure's discovery opens to a sort of differential nihilism.

### *9.3.2 The Debate on Individuation*

Structural linguistics is the most original philosophical position produced in the twentieth century. It does not only conflict with traditional ontology; it also re-opened the debate on individuation, which seemed definitively closed by Leibniz. In western philosophical tradition, "individuation" is a process that is supposed to explain the genesis of the individual starting from Universals. Let's think about a black stove–Ch. S. Peirce's famous example. The stove is black because it embodies blackness. This point of view considers a multiplicity of individual, unique elements as generated by a small amount of (positive) general qualities, the Universals. So the philosophical problem is how individuation works, starting from this small finite set. For example, Scoto proposed that "being an individual" is also a general quality (*haecceitas*), like "blackness".

Leibniz closed the debate by proposing his famous law: if $x$ and $y$ are considered equal if and only if they share every conceivable property, then no difference between entities can be considered accidental, because any difference is sufficient to distinguish them. So, there are no privileged universal substances with respect to what certain entities may be considered individual realizations. Leibniz's point fits modern sensibilities, which attributes an autonomous individual existence to singular entities. With Leibinz, we enter a universe of lonely, monadic positive entities.

Through the discovery of pure negative differences without positive terms, structuralism re-opens this debate. Positive individuals are determined by a system of pure differences. This is not a return to philosophy of the middle-ages, because the differences are *negative*. Nevertheless, in this new framework, Leibniz's law fails, as we are going to demonstrate.

### 9.3.3 Leibniz' Failure

It's time to focus on that which distinguishes structuralism's conception of the terms "system" and "structure" from the many other uses of these terms.

I will start with the notion of the identity of an element. Cultural elements do not exist in the same way as everyday objects exist. Cultural elements do not have a positive identity. If I move my television from one corner of my living room to the opposite one, it will still be a television, metaphysically speaking. Cultural elements do not behave the same way.

For example, if we compare the term "isle" to the term "island" without considering the system, we might believe that the first is a short form of the second. This is a false etymology, and the meaning is misleading. The term *isle* derives from Latin *insula* through the French *isle.* According to different reconstructions, *insula* derives from "in salo", i.e. "in the sea". If we read the American Heritage Dictionary of the English Language, we discover that the term *island* has a totally different story and meaning. It derives from Old English *īeg-land. Land* is related to the Indo-European root *lendh-*, whereas *īeg* is related to the root **akwā-*, which means *water: island = water + land.*

*Isle* and *island* share the same overall meaning, this suggests a relationship between single elements. However, the two terms should not be considered outside the system of English language. They occupied the same place in the system, so they have been (orthographically) assimilated for economy—cf. Martinet (1952).

Let's consider Leibniz' position: for any x and y, if x and y have all the same properties, then x is identical to y. Since the only two properties of a root as isl- are its sound and its meaning, structural identity turns to be non—Leibnizian: there are not differences in the sound of the two roots, and there are no differences in the referent. Nevertheless, they are not the same thing. The two roots *isl-* cannot be considered the same root and they don't have a positive identity. Linguistic roots are not televisions. We cannot shift a root from one corner of our living room to another one without changing it, because its identity is differential and positional: it depends on all of the linguistic elements to which it is linked. This means that we cannot take its identity for granted. This is a highly important epistemological feature of structuralism, as Milner noticed (2002).

### 9.3.4 Language and Quantum Logics

What kind of conclusions we can draw from the existence of two (or more) indistinguishable roots? It should be mentioned that they also represent the intensional failure of the principle of substitutivity, which is the logical version of Leibniz's law: if two terms can be mutually substituted without affecting the truth value of any proposition in which the terms occur, then the terms share the same meaning. This principle is used to study language in the analytical tradition. In my opinion, its failure demonstrates to what extent it poses an incorrect metaphysical relationship between language and reality. For example, from a synchronic point of view, the substitution of the two "isl-" preserves the truth-value in every context. We should conclude that in the system of modern English there's just one sound "isl-" with just one meaning (a tract of land completely surrounded by water, and not large enough to be called a continent). However, it is true that in the modern English system the root "isl-" derives from Latin, *and* it is true that in the modern English system the root "isl-" derives from Old English, without contradiction. If we compare the situation to a (meta) physical one, the original application field of Leibniz's principle, this would lead to an absurd analogy: there's only one TV in the living room, it is true that this TV comes from a factory in Italy, *and* it is true that this TV comes from a factory in China.

The fact that structural identity is non-Leibnizian does not mean that it is illogic. This is the point: analytical philosophy applies and superimposes simple logical systems on language. There are less ingenuous logical systems, developed to be more respectful toward the way in which different disciplines use language. Surprisingly, Simondon exemplified ontological problems regarding individuation with Quantum theory. Nowadays, the logical language developed to describe discursive formations in Quantum physics help us handle similar "identity problems". For example, in first order quantum logics, a predicate can be true of a class, without being true of any of the elements of the class. So, two quantum-sets extensionally equivalent don't necessarily share the same intensional properties. Another failure of both Leibniz's principle and the principle of substitutivity—cf. Dalla Chiara and Giuntini (2002).

### *9.3.5 Never Ending Individuation Processes*

Thanks to the new, weak structuralist notion of identity, many oppositions were discovered in language which do not seem to have a complete independent identity. An interesting example can be found in *participative oppositions*. According to Hjelmslev (1935) we have a participative opposition when an item is opposed to a category that includes it. For example, in English, the Saxon genitive carries the value of "detachment", and it is opposed to a "not-genitive" category that can mean, without distinction, "approach", "quietness", and even "detachment", when anticipated by the preposition "of".

Normally the saxon genitive is opposed to not-genitive form: Elton John's/Elton John. Nevertheless, when selected by the preposition "of", we can use the not-genitive form to express the same value carried by the saxon genitive: the son of Elton John/Elton John's son. For example, we can compare the forms: (1) "Elton John's son Zachary high-fives Bono at an Oscars party" and (2) "The son of Elton John now has an unlikely godmother—Lady Gaga" Participative oppositions constitute an interesting subterranean stream in Hjelmslev's works. He never abandoned them, we still find them in Hjelmslev (1975, def. 71) defined as "correlations between functives which share common variants".

### *9.3.6 Language Is Our Reality*

Due to the mutual dependence between the identity of an element and its system, Structuralism offers a new perspective on the issue of individuation. Since cultures are made of systems working together, it is not possible to pose an ontological dichotomy between not-leibnizian entities and Leibnizian ones. There is no distinction in principle between the world in which we find "roots", and the world in which we find "televisions", because we use the first to construct our comprehension of the second. We are never in an extra-semiotic situation, we are always inside our own cultural system. Following Simondon's train of thought, this weak notion of identity is not distinctive only for semiotics. Generally speaking, from Quantum theory to embryology, from psychology to social sciences, the individuation issue seems to characterize contemporary epistemology—cf. Simondon (2005).

## 9.4 Greimas' Actantial Syntax

As we have seen, the indirect legacy of morphogenesis in Greimas' work is represented by the link between actantial syntax and Propp's morphology, inspired by Goethe's ideas on form. It is not surprising that René Thom focused an important part of his work on actantial syntax. Following the morphogenetic approach, Thom's general goal was to substitute semiotic formalism, based on terms and relationships, with a dynamic topology of positions (places, localizations) and of co-localizations —cf. Petitot, "Catastrophe", in Greimas and Courtés (1986). This is consistent with the Gestalt/Bildung opposition. Semiotic formalism is not "wrong" in absolute terms, but it is unable to represent meaning as a transformative activity, as metamorphosis, or "in-formation", in the terms used by Simondon. For this reason, in the following paragraphs we will present Greimas' actantial syntax (Greimas 1976) in order to compare it to Thom's model, and analyze its respective strengths and weaknesses.

### *9.4.1 A Valuable Tree*

Greimas describes textual semantics as the circulation of values in a given text. Values are embodied in objects. These are disputed or simply tranfered by two or more subjects. For this reason, Greimas' actantial syntax represents syntagmatic chains of elementary utterances whose core is represented by the transformation of the initial state. The canonic form of the utterance, in Greimas' metalanguage, can be represented this way:

$$D \rightarrow (S \cap Ov) \text{ or}$$

$$D \rightarrow (S \cup Ov)$$

The symbol → represents the function "transformation"; the symbol ∩ represents the function "conjunction"; the symbol U represents the function "disjunction"; the symbol S represents a Subject; the symbol O represents an Object; the symbol v represents a semantic value carried by the object, which enters in conjunction or disjunction with the subject; and finally, the symbol D represents the condition of the junction, which is the main focus of our attention. Greimas calls it the "addresser". Simondon (2005) would most likely identify (S∩Ov) or (SUOv) with metastable states. The transition between two metastable states is always possible, the addresser expresses the condition of the transition. For example:

formula: $D \rightarrow (S \cap Ov)$; interpretation (King James 1611): "In the beginning God created the Heauen, and the Earth."

In the example, a collective Subject (Heaven and Earth) conjoins with a Valuable Object (existence); the Adresser (God) expresses the condition. Greimas' choices are close to Chomsky's generative methods. A clue in this direction comes from the lemmas "Tree" and "Bracketing" in Greimas and Courtés (1979). According to the authors, bracketing is a representation system that can be translated into the form of a tree. For example, we could represent a canonic enunciate in this manner:

This way the hierarchic relationships between the actants and the levels of articulation of the structure are easy to represent, a method which is inspired by Tesnière's stemmas—cf. Tesnière (1959). Each node in the graph represents a function, and each branch represents a functive of the function. The graph clearly shows how the addressing function → is not at the same level of the conjunction ∩, which is its hyponym in the graph. The same manner, the functive D is situated on a plane which is transcendent of the S and O plane—cf. "Addresser/Addressee" in Greimas and Courtès (1979). I have proposed a more complex generative grammar for actantial structures, in a chomskian fashion, in Galofaro (2014).

### 9.4.2 Critical Discussion

How to interpret our graph? When dealing with symbols such as S, D, Ov, we have the impression of manipulating positive units. We could even assign a symbol to the "neuter" element, such that XU = UX in a chain—Chomsky (1955), and feel as if we are dealing with a "thing". Symbols give us the impression of being positive, somehow defined once and for all. Instead, structural units are not positive: they mark differences, not things. According to structuralism, a subject is not a positive entity, but a relational one. The subject is always an element in a system that determines it. For example, a subject implies a non-subject, an anti-subject and so

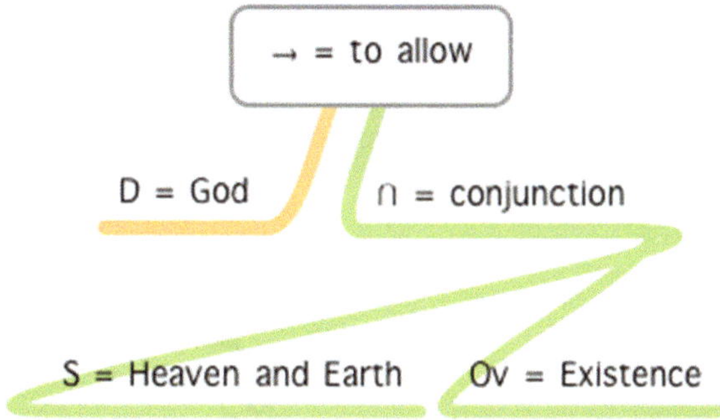


**Fig. 9.1** Tree-representation of Greimas' canonic utterance. The figure has been created using the software *Mind node lite*. The symbol ' = ' introduces the particular semantic interpretation of each syntactic node in the considered universe

on. In other terms, S is the Subject only due to a differential function that allows us recognize it because it is not an object Ov, or an Addresser D and so on. A Subject does not exist outside the functional relation in which it is involved: lonely subjects don't exist. If we look at Fig. 9.1, the nodes express this *dichotomization* function. This is consistent with the differential epistemology that distinguishes structuralism form other sciences.

Petitot considers Thom's actantial graphs as the geometrical version of Tesnière's stemmas—cf. J. Petitot, "Graphe actantiel", in Greimas and Courtés (1986). There are many reasons why we might want to propose a different formalization. The first is the *Gestalt/Bildung* opposition. Greimas' formalism describes meaning as a *Gestalt,* as the static result of a metamorphosis between two meta-stable states. It is important to create a formal model of the metamorphosis in itself, choosing between different possibilities. Such a model should address the individuation of the actants, it could respond to the issue of the constitution of the immanence plane, the abstract space in which meaning, as a form, is generated. In fact, if we don't describe the way in which the immanence plane is constituted, it seems as exoteric as hyperspace or hyperuranion.

## 9.5 René Thom's Archetypal Morphology

René Thom (1989) uses the greek word *agon* in order to label the conflict between two *actants* (a subject S1 and an anti-subject S2), which is the origin of the virtually endless circulation of an object Ov between the two. This archetype translates different situations into formal terms: a tennis game; economical exchanges; the rivalry of two lovers for the same woman—or vice versa.

In order to formalize what happens, Thom recalls the *semiotic square*—cf. Greimas (1970):

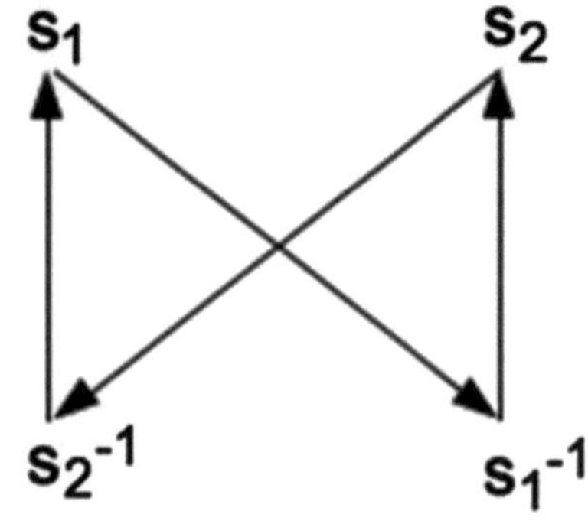


**Fig. 9.2** The semiotic square

The square in Fig. 9.2 shows the relationships between two contrary terms $s_1$ and $s_2$, and their respective negations $s_1^{-1}$ and $s_2^{-1}$. For example, if $s_1$ = /culture/, and $s_2$ = /nature/, then $s_1^{-1}$ = /not-culture/and $s_2^{-1}$ = /not-nature/. The arrows represent the circulation of the *Ov* between the four angles: during this journey, the object becomes the repository of the different values *v*, which Thom interprets as different conceptual *prégnances: not manifested forms that select the salient* O.

In order to exemplify how this works, Thom's refers to Petitot (1979). Petitot describes St. George's myth as a conflict between a Good force and an Evil one. These forces cause the circulation of an object *Ov, which* is the princess:

1. the princess leaves the city ($s_1 \rightarrow s_1^{-1}$);
2. the princess is caught by the dragon ($s_1^{-1} \rightarrow s_2$);
3. St. Georges kills the dragon and frees the princess ($s_2 \rightarrow s_2^{-1}$);
4. St. Georges returns the princess to the city ($s_2^{-1} \rightarrow s_1$).

Using Greimas' (1970) and Tesnière's (1959) terminology, Thom calls *actants* the two conflicting forces. Thom had beforehand been working on archetypal graphs to explain syntax, starting from "natural" situations that our ancestors could perceive in their natural environment—cf. Thom (1983), Wildgen (1982). By formalizing the semiotic square, Thom was also trying to explain narrative structures through the same archetypes. In particular, he notices that (1) and (3) are catastrophic changes.

### *9.5.1 From Catastrophes to Actants*

How can Catastrophe theory create a formal model for actantial schemas? Let's have a look at the *cusp* catastrophe:

In Fig. 9.3, we can see two spaces. The space on the left (W) is the control space; the space on the right (S) is the catastrophic system. A and B are the parameters (w) which define a path © in the external space W—they express the coordinates of the point we see in W. The internal states of S are determined by w: when some critical value is exceeded, S abruptly changes its internal states. Let's focus on S. The manipulation of the parameters modifies the inclination of the curve to the left or to

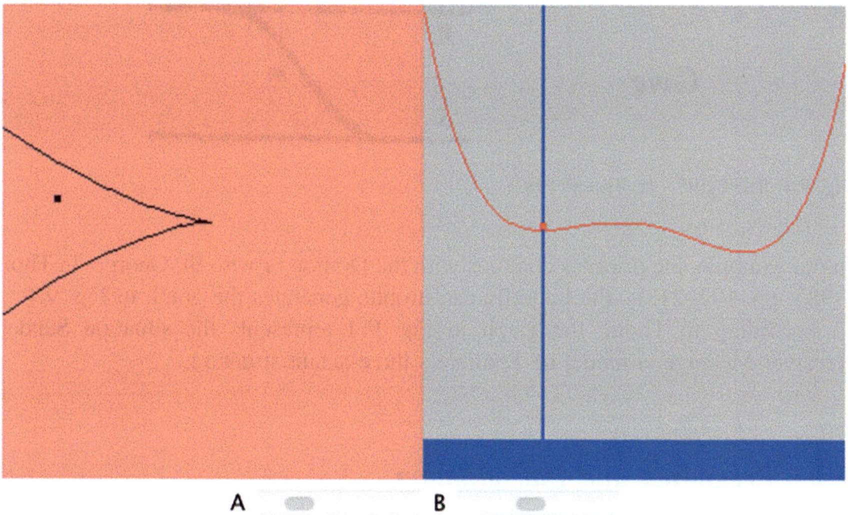


**Fig. 9.3** The cusp catastrophe, designed with catastrophe teacher (http://l.d.v.dujardin.pagesperso-orange.fr)

the right, causing the point at the intersection between the curve and the vertical line to fall to its left or to its right, and the system to reach a new meta-stable state. The critical values constitute the catastrophic set K. It is worth noticing how K categorizes the control space W, marking a border in it and distinguishing different (in our case: two) domains that correspond to the internal states of S—cf. Petitot. "Catastrophe", in Greimas and Courtés (1986).

If we interpret the internal states of an elementary catastrophe (W, Kw) as actants, and if we follow a path © in the external space W, the points of the path that traverse the catastrophic set Kw can be interpreted as interactions between actants. René Thom called these schemas *Archetypal morphologies,* and formalized the relationships between three actants: the Subject, the Object, the Anti-subject. Depending on the complexity of the catastrophe considered, we can have archetypal morphologies as the *capture* of an object by a subject (S∪O) and (S∩O); the *conflict* between a subject and an anti-subject (S/-S); or the *transfer* of an object from an anti-subject to a subject (-S → O→S)—cf. Petitot, "Morphologie archétypes", in Greimas and Courtés (1986). For example, the cusp we saw before is the basis for the *capture* archetypal morphology—cf. Thom (1983:213). A more complex catastrophe, the *butterfly,* is the basis for the *transfer* morphology.

### *9.5.2 The* Transfer *Archetypal Morphology*

Thom describes the circulation of the object between two subject as Greimas' *junction* (1983): the *conjunction* of Ov and $S_1$ implies the *disjunction* of Ov and $S_2$.

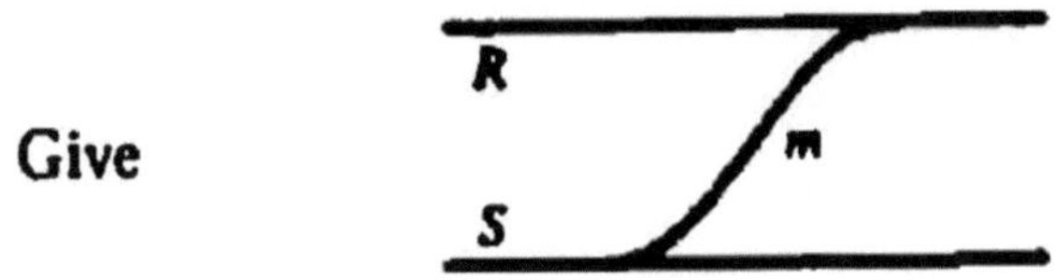


**Fig. 9.4** The "gift"—cf. Thom (1983)

In our example, the princess conjoins with the Dragon or with St. George. In Thom (1983, pp. 192–213)—the butterfly catastrophe generates the graph in Fig. 9.4:

According to Thom, the graph in Fig. 9.4 represents the situation Sender-Receiver-Message, a model of Tesniere's three-actant structure.

### *9.5.3* Prégnance *and Individuation*

In Dujardin's cusp simulation represented in Fig. 9.3 we manipulate the parameters w, which cause the transition between the states in the system S. Now we could ask: how can we interpret this "causal parameters" outside the simulation? According to Thom, the two contrasting actants (subject and anti-subject) are just the individual salient forms in which two opposite conflicting *prégnances* are invested. From the point of view of the immanentist philosophy we could say that salient forms manifest immanent *prégnances*: due to its propagative properties, a *prégnance* is never localized in space, except for the salient form in which it is invested. In my opinion, this is equivalent to saying that *prégnance* is an immanent feature. As Thom writes, catastrophes theory let us reduce the discontinuity of an observable quality (a salience) to the continuous variation of a not observable quantity (the *prégnance*).

So, salient actants are well individuated. This leads to the question: what about the individuation of the *prégnance*? According to Thom (1991), the individuation of a general *prégnance* remains obscure. He proposes considering the conflict as a possible cause, but in this case we would obtain a sort of tautology: we identify conflicts with *prégnances*, and *prégnances* with conflicts.

### *9.5.4 Participative and Privative Communication*

According to Petitot, the Catastrophic model is a junction model (conjunction, disjunction) which is a dynamic version of the primitive categorial opposition between "continuous" and "discontinuous"—cf. Petitot, "Catastrophe", in Greimas and Courtés (1986). We can once again interpret the situations in terms of *meta-stability:* the *junction* is the shift between two meta-stable states. The most stable state, irreversible, could represent the degeneration of a circular structure into a non-reversible one. Simondon (2005) exemplifies death as “stable” as opposed to the various meta-stable states of living.

But the junction represents only the static part of the canonic utterance formalized by Greimas. What can we say about the Addresser and the transformation function? Being the condition of the junction, the addresser is not present in the space K. The addresser can most likely be compared to the external control space W and its parameters. Consequently, we cannot find the addresser among the actants that characterize the archetypal morphologies. As a result of this, these morphologies are not the exhaustive counterpart of Greimas’ canonic utterances. Greimas’ addresser is more similar to Thom’s *prégnances*.

Where is the addresser in Thom’s model? At a first glance, we might recall the transfer morphology (−S→O→S) identifying the addresser with -S. Thom’s examples concern situations such as “I go from Paris to Rome” or—cf. Thom (1983, p. 182)—“Eve gives an apple to Adam”. In these cases, the first subject loses the valuable objects. We will call these relations “privative”.

Now, let’s test this model. Let’s think about the transmission of knowledge. When I teach a lesson to my students, I don’t forget the information at the end of the class. In this case, the first subject does not lose its valuable object, even if the second acquires it. In these cases, Greimas speaks about “participative communication”. In some situations, when someone transfers a value to a subject, it does not lose the value, but remains conjoint with it. Greimas’s example concerns the proxy of the power. We find the same process in the example:

In the beginning God created the Heauen, and the Earth.

“God” lets “Heaven and Earth” join with “Existence” without necessarily losing the value “Existence”. Unfortunately, Thom’s graph in Fig. 9.4 does not catch the difference between privative and participative communication. On the contrary, Greimas distinguishes the relation between the couple subject/anti-subject and the couple addresser/addressee. In the formula:

$$D \rightarrow (S \cap Ov)$$

The addresser D does not lose the value carried by the object. On the contrary, the archetypal morphology of the transfer can be formalized this way:

$$D \rightarrow (S_1 \cup Ov \cap S_2);$$

As Marsciani and Zinna (1991) notice, this is simply a concise way of notating the coupling of two formulas:

$$D \rightarrow (S_1 \cup Ov);$$

$$D \rightarrow (S_2 \cap Ov);$$

In this model, the Addresser expresses the condition that lets the subject become conjoint or disjoint from an Ov. We could say that the Addresser manipulates the parameter w of the catastrophe. For this reason, understanding how this condition works gives us a clue regarding Thom's issue with the individuation of a general *prégnance*.

### 9.5.5 Closed Versus Open Universes

In the transfer archetype, the Ov becomes disjointed from the first subject S1, and becomes conjoined to the anti-subject S2. This is correct, but only in a closed universe where values are stable and well individuated: a stationary universe. Thomian's semantic universes are closed and stationary. They are manifest finitarian spaces in which subjects, objects, and values are already present. No values are created, no values are destroyed. They all exist with a degree of certainty that equals 1. They are all positively determined as points on the curve. In this totally deterministic universe there is no uncertainty, probability, change. Actants desperately fight for a finite amount of objects and values as happens in the struggle for life. Thom has a malthusian approach to culture. Conflict is the engine, alimented by the propagation of mysterious, not individuated *prégnance*, whose source remains hidden and unexplained. According to Thom, the origin of language should be identified in "the need to inform others of a change in the surrounding milieu, of a 'phenomenological' catastrophe"—Thom (1983, p. 181). At the same time, "The competition for space is one of the most primitive biological forms of interaction. This competition then takes the typical forms of elementary catastrophes: catastrophes of capture, of giving, of excision …" Thom (1983, p. 178). The effect is the naturalization of social competition, in other words, his cultural representation through ideology. Quite the opposite, thanks to the introduction of an addresser function, Greimas's semantic universe is open. In participative communication, the addresser acts as a source of value, which puts the universe in communication with a transcendent dimension, a heterotopic axiological space. Investigating this situation could provide a clue for solving the problem of the individuation of a general *prégnance*.

### 9.5.6 Immanence and Transcendence

According to Greimas, transcendence is a feature of myths, distinguished by its inexhaustibility. It is not just a source of values, but also of valuable objects (the cornucopia; loaves and fishes) and modalities, as /power/. It is one of the most fundamental narrative dimensions. In my opinion, it could also be one of the oldest, if the morphogenetic perspective on the evolution of the narrative form is correct—cf.

Propp (1948) and (1984). Having ignored this fundamental dimension, Thom's description of the immanence plane seems flat.

Another impact of the adoption of an open model is the indeterminacy of the actantial functions. In a closed universe, for each instant $t_x$ and for each object $Ov_y$ there will be a couple $-S_z/S_z$ and a precise state of junction between $Ov_y$ and $-S_z/S_z$. The entire universe can be described through n-tuplets (S, Ov, t, x, y, z). In fact, Thom's catastrophe theory is totally deterministic; furthermore, he expressed skepticism toward quantum theory—cf. Thom (1991).

Now, the introduction of value in the semantic space creates both the two actants and the function between them. It presuppose a time, anterior to its introduction, in which the actants were not present; it presupposes also the process of their individuation. So, let's propose a little semantic universe. At the beginning ($t_o$), there are three unrelated actors. This means that the state of their junction is not determined. For this reason, we will introduce new junction values: Conjunction (1); Disjunction (0); Indetermination (?). At $t_1$, one of the actors virtually plays the actantial role of the addresser, introducing a new value in the universe and restructuring the relations between the actors. We can predict that a couple subject/anti-subject will be created: one of them will disjoin from value (0) where the second one will conjoin with value (1).

## 9.6 How to Formalize the Addressing Function

How can we correctly formalize the situation in which an addresser transfers knowledge to an addressee, without losing it? Following Greimas, we must first distinguish between actors and actants. We will consider the actors of a drama as salient forms, well individuated thanks to manifested linguistic elements such as their name, as well as a chain of co-reference terms, which constitute what Greimas called an *isotopy,* a coherent semantic layer. Nevertheless, the actantial function of the actors is not manifested. There is not necessarily a one-to-one connection between actors and their actantial function (e.g. more than one individual can play the role of the anti-subject at the same time); one actor can play different actantial functions (e.g. both the addresser and the object).

How to determine the actantial function? We will start from an observation: at the beginning of the story, the actantial function is never determined. We can say that it is not individuated, in the terms of Simondon, or, that *prégnance* is not invested in any manifested salient form. At the beginning of the story, God is not the addresser, Adam and Eve are not the addressee, the serpent is not the anti-addresser: their actantial function is undetermined, and it will be individuated only a posteriori, *given their interaction.*

During the textual development, two or more actors interact. Their actantial function becomes clearer. As the relationships between the actors reach meta-stable states during the textual development, their actantial functions are partially individuated, but never once-and-for-all. Actants will not be considered as well-individuated singularities, but as characterized by a measure of probability. If we describe the situation this way, it is clear that the deterministic instruments used by Thom to represent the actants are not capable of representing textual functions. In particular, along the textual process, there will always be moments in which the actantial functions which links two actors will be uncertain, not well determined. Now, how to deal with indetermination? Shall we renounce formalizing it? It is clear that the indeterministic processes provide, in any case, some predictable results, otherwise no semiotic analysis could be possible. As a matter of facts, Greimas's instruments precisely describe the actantial syntax starting from the end of the novel, when all the processes reach maximum stability.

### 9.6.1 Quantum Computation

First we generate an abstract structure, and then we interpret it through the discursive elements. This is clearly a reversed point of view regarding the link between structures and texts. Structures arise from phenomena that take place on textual surface, because the purpose of the former is to describe the latter. Texts are teleological constructions. That is why an analysis can only attribute narrative functions a posteriori. This also means that the individuation—in Simondon's terms (2005)—of narrative functions starts from the interactions between actors along the textual surface. When we begin reading a short story, the narrative functions attributed to actors seem initially undetermined: they become progressively clearer throughout the novel.

Any model which describes the individuation process should include indeterministic techniques to build a complete model of conversion from the fundamental, to the narrative, and then to the discursive level in Greimas' generative path, while preserving the structuralist differential notion of identity.

In Galofaro (2012), I used the diagram of a really simple quantum circuit to represent how two actors had become respectively the subject and the anti-subject during an ethnic conflict. It was a first attempt and, I'm afraid, a bit imprecise: I didn't consider the addresser role. Nevertheless, I believe that quantum computation provides an useful instrument for conceptualizing and visualizing the relationship between the three actants in a simple manner.

We use quantum computation because of its powerful ability to represent uncertainty, not because of supposed ontological similarities between meaning and particles. We can consider it as a peculiar "logic", which governs discursive formations in scientific texts of a particular kind, and which features seem useful for describing the processes of semantic stabilization.

Why quantum computation? The notion of *coherent superimposition* of contradictory states allows a better characterization of the *uncertainty* of the actantial function. For example, at the beginning God is and is not the addresser; each possibility will be associated to a certain amplitude of probability.

A second property we want to represent is the differential link between actantial functions. Actants are not positive entities: they are interdefined such that each of them presupposes its opposite (e.g. subject/antisubject; addresser/addressee). We borrow the concept of *entanglement* to represent this link: two states are entangled when they depend on each other.

A third interesting feature comes from the measurement. Each measurement performed on a quantum system destroys most of the information, leaving it in a base state. We will use this feature to represent how the uncertain actantial relationships between the actors fall at a certain point into a meta-stable state.

### *9.6.2 Differences Between Quantum and Classic Computation*

Classic computation does not seem like a good model for semantics because of two main problems. First of all, it is a deterministic model, a feature which it shares with catastrophe theory and which we already criticized. From this point of view, Quantum machines can perform tasks that classical Turing machines cannot execute. For example, they generate true random numbers, not pseudo-random ones.

The second problem of the classic Turing machine is related to binary oppositions, which do not seem able to capture every possible semantic opposition. On the contrary, a qubit is a variable whose values are not just [0, 1], like the "bit" in common computation, but can assume one of the infinite states $|\Psi>$ that results from the equation system:

$$|\Psi> \;=\alpha\,|\,0> \;+\beta\,|\,1>;$$

$$|\alpha|^2+|\beta|^2=1 \text{ (condition of normalization).}$$

$\alpha$ and $\beta$ are complex numbers: they measure a probability amplitude. Following this formulation, we will try to represent semantic values as qubits of information. This may seem strange, common sense suggests that a value exists (i.e. is present) or does not exist (i.e. is absent) in the semantic universe. However, we will use the complex probability amplitudes to represent the fact that in some textual loci the function that transfers value is uncertain. So there is only a certain probability amplitude in which the value is in conjunction or disjunction with an actant. At the beginning of the text, the value can assume just the basic state 0 (not present in the semantic universe), or 1 (already present in the semantic universe).

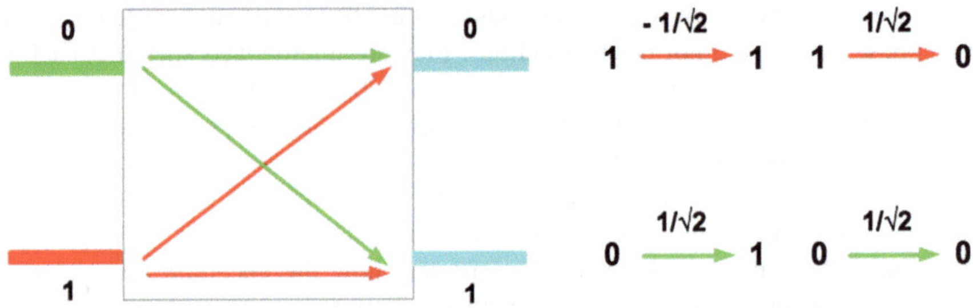


**Fig. 9.5** Action of a Hadamard gate. Given the two possible inputs (0, 1) the gate changes the value with a probability of $1/\sqrt{2}$ or $-1/\sqrt{2}$ (the sign is not relevant). Source: Lupacchini (2009)

## *9.6.3 Some Useful Quantum Gates*

In order to represent the virtual introduction of a value in our little semantic universe, we need a logic gate (the Hadamard gate H). The H-gate takes either of the basic states and places it in a superposition state, with a probability amplitude of $1/\sqrt{2}$, as in Fig. 9.5:

When the value is 1, the Hadamard gate transforms the value into a superposition of 1 and 0 with a probability amplitude of $-1/\sqrt{2}$ and $1/\sqrt{2}$. When the value is 0, the gate transforms it into a superposition of 1 and 0 with a probability amplitude of $1/\sqrt{2}$ and $1/\sqrt{2}$. Roughly speaking, the two superimposed states share the same amplitude of probability. The symbol of the Hadamard gate is represented in Fig. 9.6.

In order to represent the relationship of entanglement between our two previously separated actors, which may lead to the stabilization of an actantial relation, we need a more traditional gate: a "control-Not" gate (Fig. 9.7).

This gate links a control-qubit with a target-qubit. The gate flips the value of the target-qubit if the control-qubit is in state 1, otherwise it leaves it undisturbed. The entanglement function results from the coupling of a H and a C-Not gate. In this

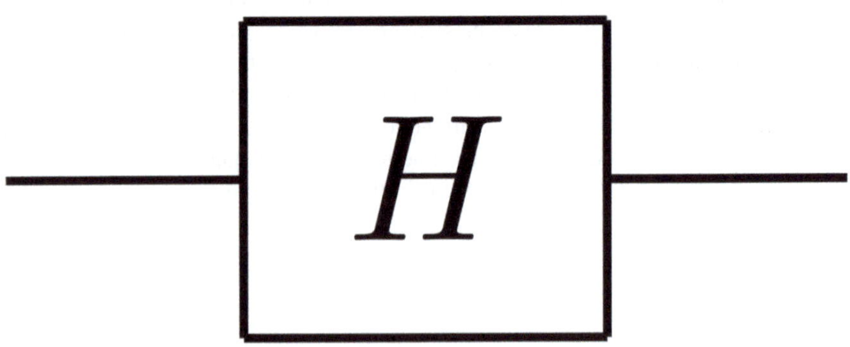


**Fig. 9.6** The symbol of the Hadamard gate

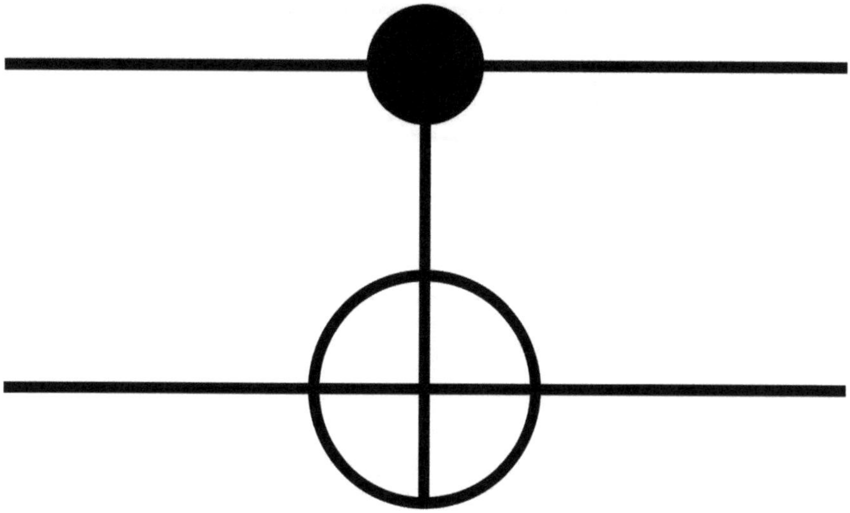

Fig. 9.7 The symbol of a control-Not gate

manner, the value of the target-qubit depends on the value of the control-qubit, which can be in a basic state *or* in a superposition of the basic states (1, 0). In the second case, we can represent an entanglement link between the two qubits.

Another useful gate for representing the exchange of value between two subjects —which is also one of Thom's archetypal morphologies—is the swap gate (Fig. 9.8):

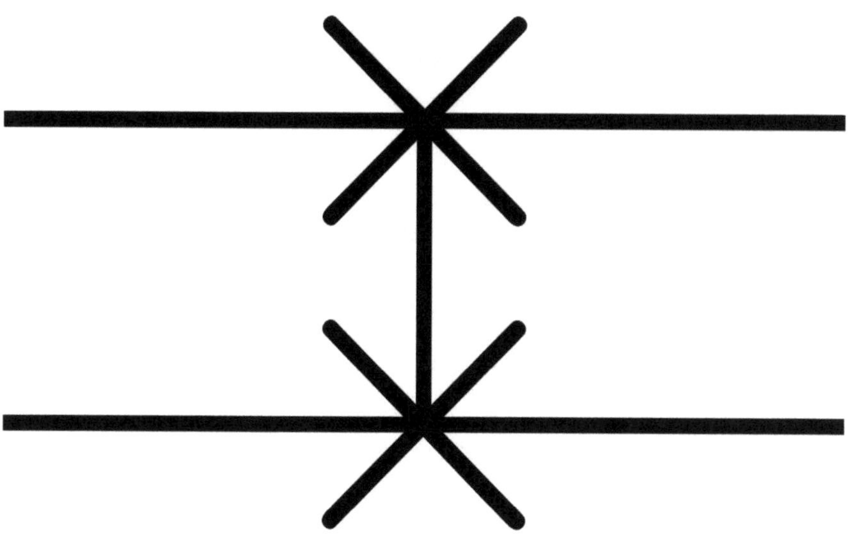

Fig. 9.8 The symbol of the Swap gate

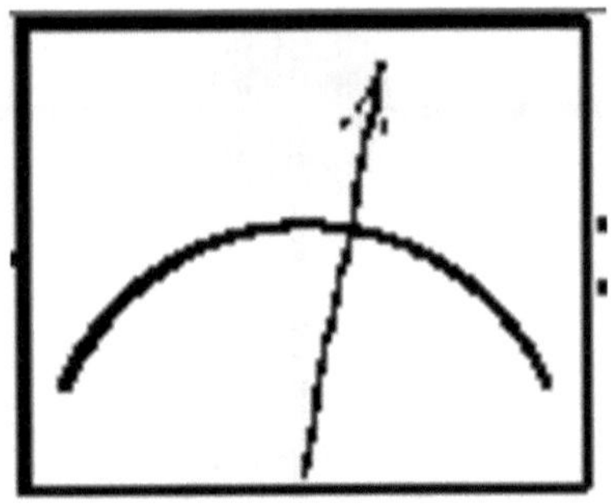

**Fig. 9.9** The symbol of the measurement gate

The swap gates simply exchanges the values of the related qubits.

Finally, in order to represent how the system reaches a meta-stable state, and the possible individuation of the actantial functions, we need a "measurement" gate (Fig. 9.9), which destroys the superposition.

Technically, the measurement is realized by calculating the *eigenvectors* of the $|\Psi>$ function.

## 9.6.4 The Circuit

Now we have all the elements to draw the quantum circuit, as in Fig. 9.10.

Each column represents a step in a process that leads the semantic value from its *virtualization* to its *realization*. Let's describe it: *in the beginning,* we have three qubits (1, 2, 3), in the state |000> with an amplitude of probability 1.0. We are trying to represent a little empty semantic universe, in which there are no values and no actantial relations.

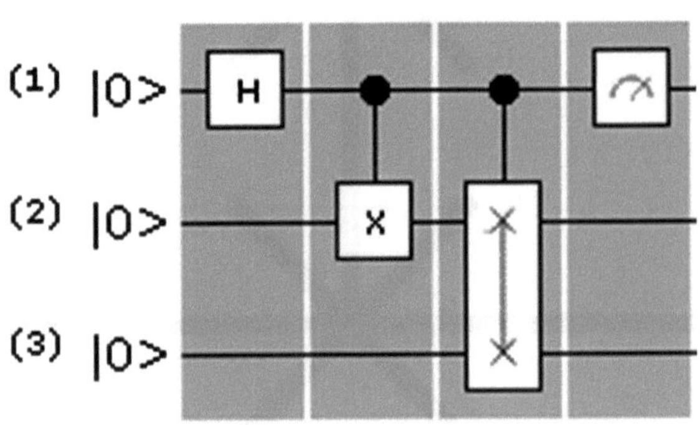


**Fig. 9.10** The addressing function. The image has been created with Zeno, universal quantum circuit simulator (http://dsc.ufcg.edu.br/~iquanta/zeno/index_en.html)

*During the first step,* the first qubit Q1 meets a H-Gate. Its value is a superimposition of the pure states (0, 1). The system can be either in the state |000> with a probability amplitude of $1/\sqrt{2}$ = 0,707, or in the state |100> with the same probability amplitude. In this manner, we would like to represent the virtual introduction of a value in the semantic universe. This value is not realized—is not "stable". Q1 will also be our virtual Addresser, which introduces value in the semantic universe.

*During the second step,* thanks to the C-Not gate, the entanglement function is established between Q1 and Q2. If Q1 is in the state |1>, then Q2 is in the state |1>, otherwise its state is |0>. But we know that the state Q1 is a superimposition of states, consequently, the system can be in the state |000> with a probability amplitude of 0,707 or in the state |110> with a probability amplitude of 0,707. Q1 and Q2 are *entangled:* if the value of Q1 becomes stable, as happens after a measurement, then the value of Q2 will become accordingly stable in relation to Q1. With this relation we are representing the virtual link between the Addresser and the Subject. Now this subject is virtually conjoint with the previously introduced semantic value.

*During the third step,* the value Q1 controls the swap of values between Q2 and Q3. As it occured in the previous step, the swap is realized only if Q1 is in the state | 1>, but we know that its state is a superimposition of 1 and 0. Consequently, the system can be in the state |000> with a probability amplitude of 0,707 or in the state |101> with a probability amplitude of 0,707. We'd like to represent, in this way, the virtual creation of the -S/S relation (i.e./-Subject/Subject/) and the virtual transfer of the object (carrying our virtual semantic value) between them (−S→O→S). These are two of the archetypal morphologies modeled by Thom and Petitot—see Sect. 9.4.1.

Finally, (*fourth step*) the measurement of Q1 forces the value to assume a pure state 0 or 1. As Q2 and Q3 are entangled with Q1, the whole system falls into one of the two meta-stable states: |000> or |101> , with a probability of 1. If steps 1–3 represent the process of the *actualization* of the semantic value, now we want to represent its *realization*: it becomes stable in the semantic universe. In the same way, the actantial relationships become stable. In the opposite case, the semantic value fails to stabilize. In a sense, it never existed, and the same can be said about the actantial relationships. In synthesis:

If we look at the last row in Table 9.1, we can write the actantial function as a simple wave function:

$$|\Psi> \ = 1/\sqrt{2}[|0;0;0> \ + |1;0;1>]$$

As a result of the measure we will obtain only one of the two vectors. In the second case |1; 0; 1>, the actantial structure is realized, in the first one, it is not.

**Table 9.1** This table represents the process that leads the semantic value from its *virtualization* to its *realization*

| Probability amplitude | States of the system | Quantum gate |
|---|---|---|
| 1 | \|000> | |
| $1/\sqrt{2} = 0.707$<br>0.707 | \|000><br>\|100> | H |
| 0.707<br>0.707 | \|000><br>\|110> | C-NOT |
| 0.707<br>0.707 | \|000><br>\|101> | SWAP |
| 1/0<br>0/1 | \|000><br>\|101> | Measure |

The data has been processed using Zeno, universal quantum circuit simulator (http://dsc.ufcg.edu.br/~iquanta/zeno/index_en.html)

### 9.6.5 What Has God Got to Do with It?

In the semantic universe of the Bible, the actantial graph represented in Fig. 9.10 can be interpreted through a process of linguistic labeling. For example:

Actor (1) = "Serpent";
Actor (2) = "Eve";
Actor (3) = "Adam";

The first Qubit will be interpreted by the label "Serpent", and will represent an Addresser D. The Hadamard gate virtually introduces the value "knowledge" into the semantic universe, embodied in the object "fruit": the probability amplitude of its presence/absence is $1/\sqrt{2} = 0.707$. The c-not gate represents an entanglement function between the (still) virtual D and "Eve", the virtual anti-subject –S—the second Qubit—that will consequently be conjoint/not-conjoint to the value "knowledge" with the same probability amplitude. The swap function is, indeed, also an entanglement function. It assures the transfer of both the object "fruit" and the value "knowledge" to "Adam", the subject S—the third Qubit—with the same probability amplitude. The entanglement functions behave as the relational identity that distinguishes the structural epistemology: in particular, S is individuated if and only if -S is individuated, where the realization of the individuation process depends on the result of the measurement.

> And the serpent said unto the woman, Ye shall not surely die: For God doth know that in the day ye eat thereof, then your eyes shall be opened, and ye shall be as gods, knowing good and evil.

What will Eve do? If the result of the collapse of the wave function is positive, Eve eats the fruit, obtains knowledge, and transfers both the fruit and knowledge to Adam. Notice how in this process she loses the fruit, not knowledge. "Serpent", "Adam", "Eve" will not only be linguistic labels. They will be coupled with the respective actantial functions, which will become semantic layers. This happens only after the process has been completed, a posteriori: the chain is a teleological one. Naturally, the formalism adopted by quantum computation does not allow us to "sneak a look at God's cards", as Giancarlo Ghirardi would say. According to different theologies, we can presume that He already knew the serpent would succeed in tempting Eve, but this does not allow us to think that the actantial function was already established from the beginning beyond a reasonable doubt. In the case of the Bible, the *bildung* process takes place in a single verse:

> And when the woman saw that the tree was good for food, and that it was pleasant to the eyes, and a tree to be desired to make one wise, she took of the fruit thereof, and did eat, and gave also unto her husband with her; and he did eat.

As a result, a new meta-stable state is reached, and a new *gestalt* is established:

> And the eyes of them both were opened, and they knew that they were naked; and they sewed fig leaves together, and made themselves aprons.

In other cases, the *bildung* process lasts for an entire novel, creating suspense and passion, or it fails, and then the virtual actantial structure is not realized. The same thing happens in our everyday life.

## 9.7 Discussion

If we look at Fig. 9.10 we get the impression that Qx is already individuated from the beginning, as a consequence of the conventional representation we adopted. This is imprecise, because at the beginning their identity was *undefined.* In fact, in a structural perspective, *identity is a relational feature.* Nevertheless, the formalism is somehow misleading because it belongs to quantum mechanics. In contemporary Quantum Field Theory, it makes little sense to talk about isolated particles, because the number of particles or states is not fixed: particles are constantly created and annihilated. In the new framework, "a qubit is never a free field [..], but it is always dressed with photons and a cloud of electron positron pairs which can freely emerge from the vacuum" (Calmet and Calmet 2012). Quantum computation is currently seen as a discrete finite case of quantum field computation where the number of qubits is finite—cf. Manohoran (2001).

### *9.7.1 The Role of the Subject*

The superposition of states can be interpreted by taking Borges' claims in his *Garden of Forking Paths* seriously: "Almost instantly, I understood: 'the garden of forking paths' was the chaotic novel; the phrase 'the various futures (not to all)' suggested to me the forking in time, not in space. A broad rereading of the work confirmed the theory. In all fictional works, each time a man is confronted with several alternatives, he chooses one and eliminates the others; in the fiction of Ts'ui Pên, he chooses– simultaneously–all of them. He creates, in this way, diverse futures, diverse times which themselves also proliferate and fork. Here, then, is the explanation of the novel's contradictions."

If we compare a novel to a wave-function, then the coherent superposition of contradictory states corresponds to the multiple virtual choices allowed by the narrative grammar. The realization of actantial structures becomes comparable to the wave function collapse, thanks to different textual devices that play the role of the "measurement". The same thing happens every time we create a model for our experience in terms of the actantial structures in our life, in the *Lebenswelt.* For example, these devices could operate constituting the actantial roles through a modalization process—cf. Greimas and Fontanille (1996), where modalization could be considered as a particular type of value transfer—cf. Marsciani and Zinna (1991); Courtés (2007)—and can be consequently modeled as in Fig. 9.10. In Galofaro (2013), the pre-individual semantization of spaces in a social house individuates the actantial roles of the lodgers. Interestingly enough, the realization is permitted by a system which influences the choices of the enunciator subject, which can't be described as an individuated autonomous subject external to the text; in fact, he or she is not distinguishable in principle from the text, being tied to the same system. This is true *a fortiori* for the everyday life set reconstructed by the observer/analyst—its *image,* as Marsciani (2012) calls it. The collapse is then somehow prescribed by the system itself, in order to: (1) be projected in a *bildung* process; and (2) to individuate a coherent narrative *gestalt.* According to Deleuze (1973) the characterization of meaning as a transcendental dimension between reality and imagination is typical of structuralism, which was interested, according to Strauss (1964) as "transcendental structures without a subject".

### 9.7.2 *The Problem of the Constitution of the Immanence Plane*

As we said previously, if we don't describe the way in which the immanence plan is constituted, it seems as exoteric as hyperspace or hyperuranion. In particular, we described the immanence plane as an abstract space in which meaning, as a form, is generated. This abstract space can be subdivided in open sub-spaces, in which it is possible to introduce new values and the subsequent stabilization of actantial functions. These are not positive, well identified once-and-for-all. They change over time with a certain probability, in relation to a shift between meta-stable states of the semantic universe. In particular, the addressing function puts different semantic spaces in communication, and introduces an order (a hierarchy) between them. Some spaces are sources of values, whereas other spaces represent their target.

The stabilization of the actantial functions is realized through the calculation of the *eigenvectors* of the function that describes a particular semantic space.

As we have emphasized many times, this semantic universe is non-deterministic, according to the epistemology of structuralism. That's why we created a model of it using instruments from quantum computation. This could seem strange, but it is not an infrequent move. As Sarti and Citti et al. (2012) wrote, the uncertainty principle has been extended from quantum mechanic to many other groups of invariance. Furthermore,

> [..] it is applied in signal processing since the work of Gabor and constitutes a central tool in the topic of coherent states and optimal measurements. It can be stated in very general terms for couples of noncommuting self-adjoint operators on Hilbert spaces and in particular we note that it has been successfully applied in image processing.

The authors apply it to the cortical model of vision. The fact that a mathematical model can be used in different fields does not prove any metaphysical similarity between them. Again, we used quantum computation because of its powerful ability to represent uncertainty, not because of supposed ontological similarities between meaning and particles. The similarities are rather epistemological. We can consider these mathematical instruments as the peculiar "logic" which governs the discursive formations in scientific texts of a particular kind, whenever we are in presence of orthogonal states that correspond to mutually exclusive features—in our case: actantial features. Quantum computation provides an instrument for conceptualizing and visualizing actantial features in a simple manner, allowing us to describe processes of semantic stabilization and handle "identity problems".

As we have described it, the immanence plane is a complex, multidimensional, and dynamic space which endlessly changes along the diachronic dimension (*bildung*); in search of meta-stable equilibrium, and in which hierarchical relations between subspaces, structures, forms, and *gestalts*, can be considered as a peculiar, transitory, and local reductions of uncertainty.

## References

Calmet, X., Calmet, J.: On a formulation of qubits in quantum field theory. Phys. Lett. **A376**, 824–826 (2012)

Caputo, C.: Hjelmslev e la semiotica. Carocci, Roma (2010)

Cassirer, E.: Structuralism in modern linguistics. In: Word, I, pp. 97–120 (1945)

Chomsky, N: The logical structure of linguistic theory, typescript, http://alpha-leonis.lids.mit.edu/chomsky/ (1955). Partially published in New York: Plenum press, 1975

Courtés, J.: La Sémiotique du langage. Armand Colin, Paris (2007)

Dalla Chiara, M.L., Giuntini, R.: Quantum logics. In: Gabbay, D.M., Guenther, F. (eds.) Handbook of Philosophical Logic, vol. 6, pp. 129–228. Kluwer Academic Publishers, Dordrecht (2002)

Deleuze, G.: A quoi reconnaît-on le structuralisme? In: Châtelet, F. (ed.) Histoire de la philosophie VIII. Le XXe siècle. Hachette, Paris (1973) (transl. 2004, "How Do We Recognize Structuralism", in Desert Islands and Other Texts, Los Angeles – New York: Semiotext(e))

Del Ninno, M.: Claude Lévi-Strauss: tra antropologia e semiotica. http://www.etnosemiotica.it/userfiles/DNM%20LS%20Tra%20antropologia%20e%20semiotica_%20Pdf%20Tahoma.pdf (2010)

De Mauro, T.: Notizie biografiche e critiche su F. De Saussure. In: Italian trans. Saussure 1916, pp. 285–363 (2001)

Ferraro, G.: Lévi-Strauss, un maestro per una strada ancora da percorrere. http://www.etnosemiotica.it/userfiles/Ferraro%20PDF%20(testo)(1).pdf (2010)

Galofaro, F.: Slavi si nasce o si diventa? La costituzione del Soggetto in 'Ritorneranno' di G. Stuparich. In: E/C, VI, 11-12, pp. 97–101 (2012)

Galofaro, F.: Social Housing e schizofrenia. In: E/C, VII, 15–16, pp. 39–42 (2013)

Galofaro, F.: Formalizing narrative structures. Glossematics, generativity, and transformational rules. In: Signata, IV, pp. 227–246 (2014)

Geymonat, L.: Storia del Pensiero Filosofico e Scientifico IV. Garzanti, Milan (1975)

Greimas, A.J.: Du Sens. Seuil, Paris (1970)

Greimas, A.J.: Maupassant. La sémiotique du texte: exercises pratiques. Seuil, Paris (1976)

Greimas, A.J.: Du Sens 2. Seuil, Paris (1983)

Greimas, A.J., Courtés, J.: Sémiotique: dictionnaire raisonné de la théorie du langage. Seuil. Paris (1979). English trans. Semiotics and language: an analytical dictionary. Indiana University Press, Indiana (1983)

Greimas, Algirdas J., Courtés, Joseph: Sémiotique: dictionnaire raisonné de la théorie du langage. Hachette, Paris (1986)

Greimas, A.J., Fontanille: Sémiotique des passions. Seuil, Paris (1996)

Hjelmslev, L.: La catégorie des cas, I. In: Acta Jurlandica, VII, 1, pp.I–XII and 1–184 (1935)

Hjelmslev, L.: Résumé of a theory of language. In: T.C.L.C. XVI, pp. 1–279 (1975)

Lévi-Strauss, C.: Le cru et le cuit. Plon, Paris (1964). English trans. The Raw and the Cooked. University of Chicago Press, Chicago (1983)

Lévi-Strauss, C.: L'Homme nu. Plon, Paris (1971). English trans. The Naked Man. Harper & Row, New York (1981)

Lévi-Strauss, C., Éribon, D.: De près et de loin. Odile Jacob, Paris (1988). English trans. Conversations with Claude Lévi-Strauss. University of Chicago Press, Chicago (1991)

Lupacchini, R.: Elementi di computazione quantistica. Clueb, Bologna (2009)

Manohoran, A.C.: The unity between quantum field computation, real computation, and quantum computation. Commun. Math. Phys. **121**, 353 (2001)

Marsciani, F.: Introdução a etnossemiótica. Centro de Pesquisas sociossemiótica, São Paulo (2012)

Marsciani, F., Zinna, A.: Elementi di semiotica generativa. Esculapio, Bologna (1991)

Martinet, A.: Function, structure, and sound change. Word **8**(1), 1–32 (1952)

Milner, J.-C.: Le périple structural: figures et paradigme. Seuil, Paris (2002)

Mondella, F.: La scienza tedesca nel periodo romantico e la Naturphilosophie. Geymonat **1975**, 225–258 (1975)

Petitot, J.: Saint-Georges : Remarques sur l'Espace Pictural. In: Sémiotique de l'Espace, pp. 95–153. Denoël-Gonthier, Paris (1979)

Propp, V.: Morfologia Skazki. Nauka, Moscow (1928). English trans. Morphology of the folktale. Texas University Press, Austin (1971)

Propp, V.: Istoricheskie korni volshebnoi skazki. Izdvo Leningradskogo Universiteta, Leningrad (1948)

Propp, V.: Theory and history of folklore. Manchester University Press, Manchester (1984)

Sarti, A., Citti, G., et. al.: An uncertainty principle underlying the functional architecture of V1. J. Physiol. **106**, 183–193 (2012) (Elsevier, Paris)

de Saussure, F.: Mémoire sur le système primitif des voyelles dans les langues indoeuropéennes. Leipzig (1879)

de Saussure, F.: Cours de linguistique générale. Payot, Paris (1916). English trans. Course in General Linguistics. La Salle, Illinois: Open Court (1983). italian trans. Corso di linguistica generale xvi edition. Laterza, Bari (2001)

de Saussure, F.: Saussure's third course of lectures on general linguistics (1910–1911) from the notebooks of Emile Constantin. Pergamon Press, Oxford (1993)
de Saussure, F.: Ecrits de linguistique générale. Gallimard, Paris (2002). English trans. Writings in General Linguistics. Oxford University Press, Oxford (2006)
Simondon, G.: L'individuation à la lumiere des notions de formes et d'information. Jérôme Millon, Grenoble (2005)
Tesnière, L.: Éléments de syntaxe structurale. Klinksieck, Paris (1959)
Thom, R.: Mathematical models of morphogenesis. Horwood, New York (1983)
Thom, R.: Cyclical structures in semiotics. In: Perron, P., Collins F. (eds.) Paris School Semiotics. John Benjamins Publishing Company, Amsterdam (1989)
Thom, R.: Morphologie du sémiotique. In: Apologie du Logos. Hachette, Paris (1991)
Wildgen, W.: Catastrophe theoretic semantics. An application and elaboration of René Thom's theory. Benjamins, Amsterdam (1982)

# Chapter 10
# Immanence/Imminence. Thinking About Immanence and Individuation

**Federico Montanari**

**Abstract** The aim of this article is to compare the concept of Immanence, or rather, the definition of "plane of Immanence" as described and invented by Deleuze (and Deleuze and Guattari) with the concept of Individuation (in Simondon, but also within a long philosophical tradition from medieval philosophy to Leibniz, as discussed also by Deleuze, as it is well known). The article will also consider the possible consequences of this investigation on semiotic research regarding: (1) the building of meaning processes; and (2) possible connections with other areas of social and human sciences (such as the Theory of Systems).

In this chapter we will try, first, to provide a definition and the origins of what Immanence and the Immanence Plane is in Deleuze, and Deleuze and Guattari's thought. Secondly, we will discuss the concept of Individuation. However, it is important to remember the importance of Simondon's research on Deleuze's philosophical investigation. In addition, in Deleuze the concept of immanence is closely linked to the concept of "expression" from Spinoza as well as his political-philosophical thought, and the way in which Deleuze and Guattari reconsider Hjelmslev's foundation of semiotics. We hope to indicate how this philosophical perspective could be further utilized as an innovative perspective shift in semiotic studies.

## 10.1 Rediscovering Expressionism

Deleuze derives the concept of immanence from Spinoza's philosophy, as is well known. In any case, this concept is very important and abundant for Deleuze thanks to its connection, in the reconstruction of Spinoza's thought (Deleuze 1968b; 1981),

F. Montanari (✉)
University of Urbino, Urbino, Italy
e-mail: federico.mont@gmail.com

F. Montanari
University of Modena-Reggio Emilia, Modena, Italy



A. Sarti et al. (eds.), *Morphogenesis and Individuation*,
Lecture Notes in Morphogenesis, DOI 10.1007/978-3-319-05101-7_10

with the concept of *expression*. This point is often overlooked in the debate surrounding the definition of immanence. Immanentism and expressionism are, according to Deleuze, two movements of thought that merge. The term "movement", as proposed by Deleuze, is to be taken literally. That is to say, something that disturbs, a hidden thought, or rather some kind of backdrop to Western thought, an alternative history of philosophy, more often submerged that emergent. It seems that the connection between immanence and the concept of expression is very important and neglected even in semiotics and the humanities. According to Deleuze, the conception of knowledge as "a kind of expression" is developed starting with Spinoza (Deleuze 1968b, p. 10). Let us see how Deleuze resumes Spinoza: "It is certain that all creatures of nature involve and express the concept of God" says Deleuze, citing Spinoza's *Ethics* (ibid.). But how can we articulate this reason? In the first place, we find in Spinoza the indication of a traditional opposition, says Deleuze, typical of a large part of the development of Western thought, between the concepts of "emanation" and "immanence". We cannot develop this point further here, but we would like to hold onto to the fundamental idea. We would like to follow this reversal of the "emanationist project" in Deleuze's work that derives from Spinoza. As we know, Emanationism is derived from a Neoplatonic, and in part Platonic, tradition, which proposed a causal principle—God, Principle, Form, essences of reality—from which the chain of beings derives its origins and meaning in the world. We can consider, just as an example, the problem of the "division of the creatures of nature" from the ninth century and John Scotus Erigena's neo-Platonic philosophy and theology. Subsequently, we find Scholastics steeped in Platonism, thinking of the One as the cause of the "participation", coming from Plato. In any case, in the first and eleventh chapter of *Spinoza and the problem of the expression* (cit.), Deleuze indicates the fundamental clue. In Plato the idea of participation of worldly beings is understood as "the power to participate". Deleuze states that Neoplatonism upsets the problem almost immediately. The Neoplatonists seem to be asking not just what happens, the effects of this participation of the cause and the divine to the multiplicity of the world. Rather they seem to attempt to discover the Principle and its effects, working on the aspect of "participated as such". Deleuze says (ibid.) that Plotinus develops a critique of Plato in this direction, emphasizing the need to study the mechanism of emanation in itself, as an "emanative" cause, instead of seeking the truth in the world, derived from an idea (or a God). As a connection: as that which is given to the person who receives a gift. Although we don't yet find the full definition of immanence here, because there is still something that "comes from the beyond", from "above" says Deleuze, nevertheless we can already see a feature of this concept, as "something that is derived from itself", which "remains itself". There is no longer a search for an external cause.

This is where the concept of immanence develops historically for Deleuze. Of course, this conception was also preceded and accompanied by the entire materialist tradition (Epicureanism, in its various aspects, and in Stoicism), as well as the intersections with Jewish (Maimonides) and Arabic philosophy, which represent one of Spinoza's sources of inspiration. And it causes Negri (1998) to highlight the idea that immanentism is a sort of "horizontal emanationism", functioning with assemblages (or chainings), a term dear to Deleuze and Guattari. In a certain manner, following Deleuze, we can say that this conception of immanence is the "great discovery" of what can be called "the modern of philosophy".[1] The break with the tradition that stated there is one world, a real, for which one necessarily had to find the transcendent external cause. This rupture, radical and blasphemous (as was considered, as is well known, Spinoza's philosophy), indicates the direction of the immanentist project. But it was also a rupture and opposition with Cartesian rationalism, against his dualism and search for a priori and abstract causative principles for clear and distinct ideas. In any case, Deleuze states that spinozan philosophy, in which the substance of the world "expresses itself", is important because Spinoza completely overthrows the scholastic and Cartesian vocabulary in this manner, that is the idea of "immanent" or "expressive" causality. Substance "is expressed first of all in itself" (Deleuze, ib. p. 145, tr. it.). Deleuze would say, resuming Spinoza "attributes are like perspectives on substance" (ibid). There can be no principle of immanence for Deleuze (yet again resuming Spinoza) without this *expressionist approach*, without the principle of expression. This is the real spinozan breakthrough according to Deleuze's thesis. But why do Deleuze and Guattari write about "memories of a spinozist" at a certain point in the final pages of *Thousands of Plateaus*, in the "glossary" of their philosophy? Do they intend to declare that Spinozan thought has been overcome in some manner? Or was it because in the mid '60 s in France there was a movement and groups of scholars (Marxists who were critical of official and dogmatic Marxism) that called themselves "spinozists", such as Althusser (cf., 1998)? Or maybe it was due to the fact that in the years immediately following *Thousands of Plateaus*, Deleuze began to work on Leibniz, who proposes a different conception of expressionism that is linked to "becoming", to the event and process dimension, and to multiplicity? Partly (but only partly), as we know, Leibniz proposes to criticize and not to accept Spinoza's thought (see, Leibniz 1999). But these do not seem like the answers. In the "spirit" of their work, both in *Thousands of Plateaus*, and later in *What is Philosophy?*, what seems to prevail is the idea of a transformative dynamic inside the concept of expression itself, which leads Deleuze and Guattari to write about "memories of a Spinozist". Memory for Deleuze is an eternal return, quoting Nietzsche, but also, as in Proust, contemplation and, at the same time, research and implementation of the concept of expression. The same thing in art, and in political and social practices (consider, for example, "post-feminist" thought, such as in Rosi Braidotti, which incorporates the concepts from Deleuze and Guattari).

[1] For a broad overview and discussion of the origins of the concept of immanence, in particular from modern and contemporary French philosophy, see Ciccarelli (2008). In addition, regarding Spinoza as the thinker who "inaugurated" and starts modern philosophy, there are interesting considerations in the introduction of Seymour (1991), in *Ethics*, in which, also resuming Wolfson (1934), he insists on the fact that the idea of "modern" should not be thought of in a chronological sense, but as a true *break* with a tradition. Spinoza breaks with a religious tradition. Modern thought begins when philosophy ceases to "draw its own information" from religious traditions. Whereas in Descartes there is not yet this break, despite his intention of founding a new method. Many scholars, Feldman (cit.) insists, support the thesis that Descartes is still "medieval". See also Alquié (1981).

## 10.2 Early Origins of the Plane of Immanence: Modes, Assemblages, Substance

In this regard, we would like to try to outline the relationship that takes us from Spinoza to semiotics, through Deleuze. Meanwhile we can recall the very modern conception concerning the issue of "attributes", from Spinoza to Deleuze's reading. These would act like real "operators" (proto-observers), immanent to the very substance of the world. Attributes, by being placed in the substance of the world, and at the same time observing it, allow for the same knowledge. "Attributes", says Spinoza, "are the eyes of the mind with which we perceive". We find here, we believe, an anticipation of a more radical immanent conception of knowledge and perception. That is to say, perception and knowledge are potentially already installed in the world (and in discourse, we would say today), but that, in installing themselves, build the perception and knowledge of the world. The attributes are what the intellect perceives of the substance. Here, in fact, is the well-known definition of attribute, "By attribute I understand what the intellect perceives of a substance, as constituting its essence" (*Ethics*, Part1, D 4). As Banfi points out (ibid., p. 214–215; see Spinoza, Ethics, II, 37), if spinozan "common notions" form a sort of fundamental intermediate level—rational knowledge—between sensitive knowledge and the third highest rank of general and intuitive knowledge, we must recall that they are generalizations derived from experience, but are not "mere abstractions", as is instead the case in the Aristotelian tradition. They are the construction of a "plot of infinite ways", and at the same time, a resolution of this same plot. We should remember, in this regard, that according to Deleuze (and Guattari), constructivism and expressionism accompany each other as philosophical attitudes (see, 1991, en. tr., p. 35). For Spinoza, in the end (see Banfi, ib.), the common notions are based on the existence of an infinite number of ways. However, they also define the "schema of the finite", made of concrete existence, direct connection with the substance of the world comes from them. This schema is articulated through attributes, such as the manifestations of the substance. But precisely because of this, says Banfi, this schema is able to give rise to the construction of additional possible "lines of connection". Here, once again, this spinozan immanent auto-productive dimension appears. Addressing the definition of the attribute, Banfi (ib.) indicates the clearest example: we know only two attributes of substance, of all of those possible, the "extension" and "thought". In any case, the attribute "extension" may correspond in the "physical" and "concrete" world, for example, to "movement". However, regarding ideas, it could correspond to the "infinite movement of knowledge", just like vision of a movement of connections that crosses through us. More generally, in the words of the Feldman's introduction in *Ethics*.

> By ascribing extension to God, Spinoza was able to overcome the dualist dilemmas that plagued Aristotelian and Cartesian metaphysics. (…) For Spinoza there is a fundamental continuity between the ultimate cause of everything, God, or substance, and that which is caused, or the *modes*. This continuity is an expression and consequence of God's causal

activity. Spinoza's God is an 'effacious' being that is incessantly productive; for, 'from the necessity of divine nature there must follow an infinite number of things in infinite ways' (Prop. 16, Book I). The absolutely infinite being God is infinitely causative; hence, the world is 'maximal rich': it contains everything that can possibly exist. (Feldman, Introduction, Spinoza, *The Ethics*, p. 10).

## 10.3 Modal Existences and Intensity

Spinoza scholars (see Banfi 1969) and, above all Deleuze (ib.) insist on this important point: *Ethics* begins with the idea of an "unconditional cause" (*causa sui*), the substance is one and one only and is the cause of itself. This god-nature, substance, in Spinoza has an absolute power of existence, but at the same time of ideas. It thinks itself, and understands itself. It includes everything that is produced (see Duffy 2006). However, right away Spinoza defines this immanence as that which "expresses itself". It organizes itself into attributes and modes, articulating itself. This strength of immanent self-organization, as anticipated by Deleuze and other scholars (see Duffy, ibid.), represents a great innovation in Spinoza's thought. Anticipatory even of our contemporary epistemology (think, for example, of the theory of auto-poietic systems, particularly Luhmann's theory of social systems; see 1990). Duffy insists, among other things, on the fact that the concept of intensity (qualitative variation; see also, on this point, Deleuze 1968a) is in sharp contrast, as emphasized above, with that of Hegel, another "reader" of Spinoza. For Spinoza, things and the world are nothing but affections (modes) of the attributes of substance, but are not derived from it. They are themselves substance. They are, in fact, their articulations. There is no existence outside of the attributes. In this sense the attributes express the substance itself. Therefore for Deleuze this is Spinoza's first absolute innovation, in his overthrow of scholasticism. The idea that substance "expresses" and attributes are the expressions of the substance. The essence (that which is expressed) can not exist outside of the expression itself. In this regard, the fundamental point is no longer in the definition of "the order of being" or of beings, as it had been for philosophy and scholastic theology. The question becomes the "existing", or rather the mode and the power of existence. In his lectures,[2] Deleuze insists on this fascinating and central point: we are faced with an *assemblage of the "existing"*, whose existence has a "modal" character. We, the "existing", are "modal" beings as well as "ethical". Me and you, as individuals, or as social groups, or animals, but also things (for very different degrees of complexity and relationship); we all exist in as much as we express an ability to be and do (or not do). Deleuze says, "What can that animal do or not do? What is it capable of? (can it fly, eats grass, kill, etc.)." "What you are able to do?"; or "that guy there, maybe a

[2] See, for a partial collection and translation of the lessons of Deleuze on Spinoza, Deleuze (2007). Recordings and transcripts of his lectures can still be found online at: http://www2.univ-paris8.fr/deleuze/article.php3?id_article=91 (last online consultation in March 2014).

friend, what is he capable of doing now?". How far can we take it? (Consider, also, a politician, or a dictator, etc.). Deleuze insists, as we said, on this point because, utilizing Spinoza, moves from a classification (physical and moral) of the essences (Aristotelian conception and later Scholasticism) to a description of geometric diagrams, diagram of forces, powers, and the lives of beings. "Quality of powers that distinguish one being from another". We are in the midst of a *relational and differential/modal* conception of physical, psychological and social lives (although, contrary to Leibniz, Spinoza does not know, let alone invent, differential and infinitesimal calculus, but envisages a qualitative version relating to actions and passions (cf., also Cristofolini 2008). In any case, the innovative point, highlighted by Deleuze's comment and lessons is found in this vision "of modal variations", statements that are not only judgments but are assemblages of perception and bodies themselves. These assemblages occur at different speeds and intensities; different and differentiated. The modal principle is not only found in statements ("I would like"; "I wanted to"; "I wanted that…"; or, "we believed…") but in their perceptual and cognitive assemblages and variations. We find again: "an animal can or can not, do"; "that man, or that woman, what are they capable of?". Deleuze emphasizes this very point in his lectures: the formation of a dual parameter, qualitative (intensity variations) and quantitative (even if, actually, it is a type of quantity indicated by accumulative variation, shall we say, not numeric quantity but, once again, qualitative). That is to say that we will have "quantitative variations" (more power or less power) and "qualitative distinctions" (different modes of existence). This point brings us closer to our issue of individuation.

## 10.4 Events that Populate the World of Immanence: Toward a New Concept of Individuation

Deleuze, in his lectures on Spinoza, insists that the German idealist philosophers (notably Fichte and Schelling) who returned to spinozan philosophy starting from romanticism, propose a new definition of individuation, due to this conception. The traditional definition of individuation, which runs through the history of philosophy (in particular the medieval period), states that an individual is already given, and from this individual characteristics are found that define it as such. The individual is already constituted before individuation, thus becoming mere "identification" and a list of traits. Instead the definition that opposes it, coming from this reading of Spinoza, is relational and quantitative. We are no longer saying what "is" an individual, but what it "can" and "can not" do, or be. We will resume this theme later, regarding Simondon. But it is important to mention here that such a reading stresses, as is well known, a concrete relationship, de facto, thanks to Deleuze's mediation, between Spinoza's thought and Simondon's (relationship, partial and controversial, questionable and to be discussed); as well as the "different" path of the medieval philosophy, following Duns Scotus's monism, with his definition of individual and individuation as "*haecceitas*". If some scholars (see, for example, Ciccarelli 2008) have emphasized the "condemnation" and criticism of Spinoza's thought[3] in Simondon, we must here, however, think about the issue of a new definition of individuation as a process and assemblage. In this we find the link to the concept of haecceity. In this case, we are not talking about—and the scholars of Duns Scotus stress this point (see, i.e., Cross 2010; as well as Deleuze and Guattari 1980)—a "general" quality (not a *quid* or a qualitative trait), or a trait that distinguishes a "thing" or an individual from another. We are referring to the emergence of a "*this*". From a singular entity, a type of singular difference, which produces and gives rise to individuation. We no longer, therefore, think of the individual and the individuation as modes of a subject already formed, but on the contrary, these modes constitute the *field of the construction of subjectivity*, both singular and collective (see also, Matheron 1969). In a kind of variations, of modal "fluctuation" dances that give life to immanence itself as well as to the "pure transcendental field" (definition of immanence, for as was highlighted in one of Deleuze's last writings, 1997) punctuated by singular events.[4] Hence features similar to Spinoza emerge: the modal assemblages and chainings that define a being carve and delimit its own individuality. Certainly, in this regard, many have pointed out the contrast with another great "baroque" philosopher, Leibniz, also object of Deleuze's passionate study (see 1988), as is well known. However, even in this case, if Leibniz in different moments of his philosophical journey attacked and criticised Spinoza, we can also note his ambivalent behaviour of attraction and repulsion toward Spinoza, from "refutation" to recognition. One could say, as some scholars have (see De Gaudemar 1999), that to Spinoza's modal dimension Leibniz counters with the plurality of the monads, to the uniqueness of the substance with the multiplicity of the event. Leibniz also responds with the same definition of haecceity as a "sensation and moment in a time and place", which can participate in the principle of individuation (see, in Ciccarelli, ib., p. 152). Nevertheless Leibniz sometimes seems to condemn in Spinoza that which seems to be "dangerously close" (see De Gaudemar,

[3] Simondon classically accuses Spinoza of pantheism and, as pointed out by Ciccarelli (2008, p. 430), as well as a type of "oppressive valorisation" of being immanent for which "nothing on earth is free". (See, for an analysis of the influence of Simondon's thought on Deleuze, Sauvagnargues 2005).

[4] See Deleuze (1995). For a detailed discussion of this point (also considering Leibniz's definition of individuation and haecceity) and the definition of immanence and the plane of immanence in Deleuze and Guattari, see Ciccarelli (2008), in particular, pp. 151–153. On this point, see also Galofaro's essay in this book, with references to Leibniz's law, as well as the treatment of the issue of haecceity by Peirce. For further discussion of the concept of haecceity in Duns Scotus, and Leibniz's position, see Cross (2010). Deleuze insists on an idea that we will find more than in Peirce, in another great pragmatist such as William James, and other American philosophers such as Whitehead, and even before that in an empiricist such as Hume, and Leibniz. If spinozan modes and substance are immanence, the haecceity are a kind of occurrence, of singularities, within immanence itself (see, for a summary of this point, Ciccarelli, ibid, pp. 685-686, who points out that this idea was already present in *Difference and Repetition*).

cit., p. 76).[5] In any case, what Deleuze uses and connects in Spinoza and Leibniz (and in the end we will see also with Simondon) is "the event, inside the immanent dimension, understood as a passage, and between one haecceity and another, between one environment and another" (Ciccarelli, ib.). This passage between environments constitutes the same immanent dimension. More generally, it is also in this sense that Duffy (ibid.), and other authors emphasize that in Spinoza there is, however, the rejection of an "underlying essence" (which instead is found in Hegel's reading of Spinoza, transforming this "auto-cause", or expressive substance as the cause of itself logical auto-determination). The "engine", or the "force", is no longer "outside", or in another place, or in the subject (the "other" subject, abstract, in Hegel's reading, or of the transcendental subject). It is in a force, a power (*conatus*, as it is known in the spinozan definition). Nevertheless, the point is that the modal power is *expressed* in the degrees of intensity, it appears in the same intensive paradigm/parameter of expression.

Duffy writes (2006: 240–241), in this regard: "It is according to the logic of expression, that the variations of a finite existing mode's degree of power, of power to act, are determined by the dynamic nature of the relations in which it is further differenciated [...]. This dynamism, which determines the variations of a mode's power to act and therefore of 'what is expressed', is actualized in the concept of intensity".

## 10.5 Expressionism as an Auto-Productive Dynamic. from Morality to Ethics

What interests us most here is, once again, the importance of the rediscovery of this "expressionism" in philosophy starting with Spinoza. This discovery is the force, Deleuze emphasizes, of Spinoza. Let us try to understand better what this expressionism is, and why it is so important to the concept of immanence. What is it? Are we dealing with some sort of principle of force, a kind of *energy*, that conception that was so dear to the Romanticism, who effectively declare themselves spinozian, as well as the first theoretical linguistics that were somehow inspired by that movement (such as Humboldt)? We believe that it is something else. We are dealing with dynamics, yes, but, in fact, as Duffy points out, dynamics of variations in the intensity of modes of action and being affected. The modal nature meets the intensive dimension here. It is worth noting here, albeit briefly, a few well known points regarding Spinoza and his philosophy. Spinozan fortune manifests itself, as

[5] In Leibniz's notes, edited and introduced by De Gaudemar (1999), *Réfutation inédite de Spinoza* (the title of Leibniz's notes on Wachter's book on the Kabbalah, was invented later and added in the nineteenth century by an anti-Spinozist, defender of the Catholic faith, Count Foucher de Careil, almost, as he wrote, as "antidote" to spinozan "poison"), Leibniz concept that criticizes Spinoza for having consider not only one type of substance, but only one type of process ("emanative"), is important, as is the rejection of spinozan immanence.

we know, at various times, albeit often in a controversial and paradoxical manner. According to Yovel (1989), for a long period Spinoza's thought was excluded from the "respectable circles" of philosophy and, even today, despite being studied as a classic of modern thought, Spinoza's philosophy is sometimes considered by a certain "establishment" philosophy as "outsider" (see also, Israel 2001). In any case, we know how this reputation is characterized by political-ethical and religious implications. The moments of fortune that can provoke controversy are well-known. From the very beginning, in the Netherlands Spinoza was banished from the synagogue and at the same time was considered a dangerous atheist and was attacked by theologians of various denominations. He was the victim of an assassination attempt and forced to suspend the publication of *Ethics*. His *Theologico-Political Treatise*, even though published anonymously, was branded as "the work of an atheist Jew". All this, in the context of extraordinary cultural and economic innovation, as well as the political and religious crisis of the United Provinces (see Balibar 1985; Nadler 1999, 2001; Deleuze 1981; as well as Huizinga's work). Only a few years after the death of the philosopher (see Assman 1997), his philosophy began to circulate underground, among Toland and English philosophers, natural scientists of the Royal Society, but also in cultural environments that some scholars (Banfi 1969) have defined as pre-romantic. This moment, in anticipation of the controversy regarding atheism (opened by Jacobi), posed the issue of the relationship between nature, man and thought, the relationship between "One and All". Particularly Assman (cited above) insists on this point, regarding the "spinozian cult" (between the French neo-epicurean and libertine thought, and english deists). In brief, the central concepts of spinozism, in regards to our consideration of the concept of "immanence", "immanentistic expressionism", and after of "plane of immanence" are the following (we return here to the synthesis proposed by Deleuze 1968b, 1981, and others, see, in particular, Banfi 1969).

The first fundamental concept is, we repeat, immanence as "auto-production". All is one, one-everything. God, or, in other words, Nature, *Hen Kai Pan, deus sive natura*. This is the view that will be presented to the followers of Spinozism. For Assman (cit.), rather than pantheism we are dealing with a form of "cosmo-theism" (see Assman, cit.). We no longer think that "God is everywhere", but that "the world, nature are themselves God". Here the concept of radical immanence is presented to the greatest extent. What "has to be", in Banfi's words (ib.) regarding immanence in Spinoza. Or, following Yovel (cit.), the principle of immanence sees worldly existence as the actuality of the only being possible and as the source of ethical value and political power. Transcendence represents some sort of hardening, a conquest of vertical control, the construction of an abstract, external or absolute point of view. For Spinoza and his followers the idea of immanence implies only itself, its *existence*, as well as the methods and *strategies that it has to express itself.*

The second concept is that which is correlated to power. What is the connection? We already saw it earlier in part through the concept of modal existence. Certainly, normally one insists on the issue of corporeality (hence the issue of the relationship between immanence and affection, two places well known to spinozian interpretation). How can we specify our problem better? The real question (and very famous) that, according to scholars and in particular Deleuze, Spinoza poses is the following: we do not know "what a body can do", nor the ideas that the mind creates, of a body. What does this mean? That for Spinoza the body is not a simple passive container, nor an environment that simply serves to protect and nourish the mind, but, on the contrary, is an active transmitter and generator. In addition, Spinoza emphasizes, we only know our body. Or rather, "Man does not know himself except through the affections of their body and their ideas" (*Ethics*, Part III, prop. LIII dim.). The interpretations given by scholars (think especially of the two mentioned above, very different but united by the issue of affect, such as Banfi and Deleuze, but also Duffy, cit.) tend to converge with the interpretation of Spinoza by a neurophysiologist such as Damasio (2003). We deal with a movement of knowledge, certainly, and of the affirmation of one's own corporeal individuality, who "wants to survive". Yet also, returning to ethical issues, there are also movements of different connections between me, us and others: within the community (of which I am a living part), and humanity and other "existing" (today, regarding Spinoza, we would also talk about different cultural forms).

Here another interesting element enters, the direction towards an immanent ethics, an ethology of bodies and desires. We will see shortly that it will be Deleuze who discovers, even in this case (right in Simondon) this dimension of "ethical-ethology". But it is also Deleuze (1968b, cit.) who grasps and emphasizes, as we said, the other key aspect of spinozan ethics. Along with immanence we find the "expressive" dimension, expressionist of ethics. In Spinoza "understanding" (understood both as the ability to understand and feeling) is "*explicatio*" and "*expressio*". Unfolding as internal forms of causality, yet not intended as a priori schemas. Rather, we are dealing with the connection, and the outcome of Spinoza's other well-known concept: parallelism (as it was defined in reality later in Leibniz). There is no opposition between the body and the mind (the mind as "*idea corporis*", and the body as a "*idea materiale mentis*"), there is instead some sort of active mirroring of one another. In Spinoza there is the idea that physical and mental phenomena (extension and thought) occur in parallel, without there being a direct causal relationship. As stated in *Ethics*, "The order and connection of ideas is the same as the order and connection of things." (Part 2 P. 7).

Thanks to Deleuze's reading, we also find here the emergence and the underlining of another innovative step. We find not only a theory of knowledge but, as we have seen, at the same time, a theory of affect and passions as assemblages, as chainings: parallelism and assemblage (concepts that renews the Aristotelian tradition, regarding the effects dependent on causes, as Deleuze says, 1968b, ib.). We now find ourselves, according to Spinoza, bit by bit, faced with assemblages, chainings of ideas; ideas that are assembled with each other, but in an appropriate way. That is, at the same time expressing and implying their cause. For Deleuze we are, also in this case, dealing with a properly "expressionist" idea: the appropriate idea is the one which expresses its cause. Here then, in order to understand the effect is to extract the inherent cause (synthetic and reflective method, according to Deleuze's reading, cit.). At the same time, the ways of knowledge become like modifications, changes taking place. Here again, "in parallel", we therefore also find the assemblages of the states and affections of the body. Deployment of forms of causality, of course, but that are never seen as a priori. And with parallelism between the body and the mind, and active mirroring of one another, the novelty consists in this: in the assemblages in which we find the identity of connections between the two series (body and mind), that is, equal dignity for extension and thought (between body and soul). The order of the passions and actions exists in both body and soul, and are simultaneous orders. For Deleuze (1981) we are talking about parallelism, as well as isonomy and expressionism, and "isology" (identity of being). In this sense, according to Deleuze's reading of Spinoza, there is thought and action in the body in the same manner as thought and action in the mind (thinking of the body and thinking of the mind, making the body and making the mind). Against a conception of thought, knowledge and ethics based on "isolated signs", another conception emerges in Spinoza that is based on practices of assemblages: which we might call, today, as narrative forms and structures, starting from definition of affection and perception of the body (see also, Guérin 1992). The story of bodies is also an ethics of encounters between bodies.

## 10.6 The Second Step Towards the Construction of the Plane of Immanence: Stratifications and Crossovers

Beyond the idea of immanence in general, "the point of no return" in Deleuze and Guattari's thought comes from the elaboration of the definition of the plane of immanence. How did they arrive at this step? This seems to be really a collective product. A combination of Deleuze's research on Spinoza with Guattari's social, political, semio-linguistic, and psychoanalytic-critical work (see Dosse 2007; also Montanari 2012). In which way this conception emerges? Foreshadowed in *Anti-Oedipus*, it seems that a first image of the plane can be found in *Mille Plateaux* when at a certain point, after re-addressing the idea of the rhizome with various "axioms" (in a style that seems almost to follow Spinoza's *Ethics*) they accomplish something that is at once discursive and practical. They distribute the conceptual work on different planes (the title of the book itself) and absorb the concept of layering (strata). A piece seems to lie in this image, starting from the previous definition of what was called by Artaud the *Body without Organs* (*BwOs*). In response to the question of what this is, Deleuze and Guattari come to define it not as a concept, but rather "a practice, a set of practices", they state "but one is already on it", and, "like someone full of lice", fumbling like "a blind-man running like a crazy", "desert traveller and steppe nomad". One is already *on the plane*, as a kind of, we could say, cloth, empty bag, which the traveller carries. Why this figure? Because it anticipates the dimension of the Plane of immanence. Recall that, subsequently, in *What is Philosophy?* (1991), they affirm that the plane of immanence is not a concept but, in fact, an image of the idea. In addition, of course, already in

*Mille Plateaux*, (1980) (*Thousands of Plateaus, ThP*), we find another figure that is clearly that of the plane. But where do these other figures of a desert, and a traveller come from? And therefore, characters, figures, containers and spaces? We find ourselves faced with the foreshadowing of what will happen later, in fact, in *What is Philosophy?*. Conceptual characters that operate in space, and on the edges of the real ("*nappes*", strips, or "edges") that serve to delimit, contain, or even to miss something. In reality we are talking about the same figure, seen in its "becoming", in the dynamic of its transformation. Artaud's concept of "Body without Organs" (*BwOs*) tends to de-stratify itself, and become its own border and edge (that traveller's sack…), "all lead away from the plane of organization to the plane of consistency" (*ThP*, p. 279, en. tr.). The plane of immanence is not a space or a concept, it is actually an edge, an image (in the literal sense of the term), or better yet, an "environment" (for the construction of concepts, or of practices, action, passion, and life, as Deleuze would state in his last work, *L'immanence: une vie*, cit., 1995; see also, Deleuze 1993).

Therefore, we can better address the other more well known definition of the plane of immanence that will be given later in *What is Philosophy?*. The plane is a horizon for concepts and practices. And it is made of waves, say Deleuze and Guattari, like the water's surface, or a space. It is inhabited and paved with concepts but, in fact is not a concept or an object; "it is essential not to confuse the plane of immanence and the concepts that occupy it. (…) But it only receives a concept if we determine its components (…). Movements or elements of the plane, therefore, will seem to be only nominal definitions in relation to concepts so long as we disregard the difference in nature between plane and concepts. But in reality, elements of the plane are *diagrammatic features*, whereas concepts are *intensive features*". (ib., en. tr., p 39). That is to say, that in the plane there are relational components (diagrammatic, distinctive), and intensive traits that characterize them and inhabit them (which in turn give rise to the concepts, recouping the spinozan idea seen earlier). We had said, moreover, that the plane is therefore the environment and the image-environment of a thought. The plane, in addition, for this reason allows for movements, displacements and subsequent positions. It is therefore conceivable as a movement and wave of thoughts (and therefore with curvatures). "The plane, for this reason, moves and changes from period to period" (we would say, from culture to culture) (Deleuze and Guattari 1991, en. tr., pp. 38–39). In the end, particularly for these characteristics, the plane always implies that there are other planes. But we "inhabit" a plane in any case. The plane of immanence must therefore be thought of as a kind of conceptual "events horizon", "*l'horizon des événements possibles*" (ib.). We could think of other planes, such as the virtual dimension, which could be gradually actualized by the same events (think of what was said earlier regarding the definition of haecceities), events seen as an opportunity to move from one plane to another. There is another well known important point that brings us to the issue of the study of the processes of signification. In the reading we have proposed, starting from Deleuze's text on *Spinoza and the issue of expression*, the possibility of the reversibility of "express" and "expression" is revealed. The semiotician here finds something familiar. A substance, which is re-articulated into two planes, "of the expressed" and the level of expression. In these definitions the transition from the general concept of immanence to the plane of immanence is foreshadowed, as it is in Delueze's work with Guattari, in *the Anti-Oedipus*, in his university lectures from the 80 s, and especially in the culmination of *Thousands of Plateaus*. We can say that this transition takes place, according to our hypothesis, particularly thanks to the encounter with Semiotics. Deleuze and Guattari (thanks, in this case Guattari; see, Dosse[6]; Fabbri 1997; Montanari 2012) find in the definition of semiotic system that Hjelmslev's semiotics offers – plane of expression and plane of mutually interconnected and in reciprocal presupposition content – a way to enhance the analysis of the immanent dimension. It is not a coincidence that (the story is well known, but important to remember) Deleuze and Guattari would interpret Hjelmslev (certainly, partly unorthodox, but at the same time rigorous; see, Bondi 2011), calling him "the dark spinozan prince" (see Fabbri's comments 1997). In reality, we do not know if Hjelmslev read Spinoza (Hjelmslev experts have doubts). In any case, we believe that, starting from Deleuze's work (and Guattari's, in this case), this interpretation can be assessed as plausible as well as stimulating for further research. We will try to further clarify these points regarding semiotics by using Spinoza, and then eventually return to the relationship between Simondon and Deleuze. Substance is, for Spinoza and Deleuze (ibid.), that which "expresses itself", and attributes are "expressions" of the substance, and essence is "what is expressed" and does not exist outside of expression. Deleuze and Guattari, beginning with Hjelmslev's work (1943 (1968)), insist not only on the idea of reciprocal presupposition, but return to the issue by addressing the fundamental concept of "layering", again from Hjelmslev (see Bondi, cit.). To this concept of layers, strata, that compose the two planes of semiotic systems (forms, substances) for Hjelmlsev,[7] they add a "infra-layer" dynamic component (de-layering and re-layering, de-stratification, re-stratification). The layerings (with a figure, this time, geological, of "fault lines") decompose, and recompose between themselves, pressing on each other, giving rise to continuous new re-stratifications, movements and hardenings. Perhaps, this dimension of continuous and dynamic deconstruction

[6] For a discussion of the consequences of this meeting, and the influences that particularly Deleuze and Guattari's thought would have on semiotics, see Fabbri, 1997; Bondì (2011); and Montanari (2012); see also, Dawkins, 2005.

[7] Deleuze and Guattari's reference especially regards Hjelmslev's *Prolegomena* (1943), but clearly, with regard to the concept of layers, we can not ignore his paper *The stratification of language* (1954, in: Hjelmslev 1971). For a detailed discussion of this point we refer again to Bondi (2011), which treats the concept of the "use" in Hjelmslev, bringing it closer to Saussure's recently discovered works (the unpublished manuscripts). Bondì writes (ibid, p. 186): "Through Hjelmslev we can say, in conclusion, that any deep shift in *language* passes through its surface of use and words" (our tr.). He points out, citing Zinna: "Any deep changes pass through the surface. In the speech act there is the syntagmatic stratification of language, but also the stratification of fragments of all languages of the past from which their strata come" (my en. tr.): *"tous les changements en profondeur passent par la surface. Dans l'instance de la parole il y a la stratification syntagmatique d'une langue, mais aussi la stratification des fragments de toutes les langues du passé qui réalisent leurs strates"*.

and reconstruction of the semiotic layers seemed to be lacking in Hjelmlsev and in the model of contemporary semiotics, at least in part (cf., Caputo 2010; Ouellet 1992)? Or perhaps this model should be revisiting in his reading, or even integrated and pushed in this direction. In this regard, there are scholars (see Bondi, cit.), that, thanks to an in-depth analysis of Hjelmslev's work seem to find in the founding father of modern semiotics, in this conception of layers, the dynamic subtexts that foreshadow such a vision. Moreover, Hjelmslev in *La stratification du langage* (1954), stressed the issue of thinking about the levels that make up semiotic substance (for instance the level he called "collective appreciation", the "socio-biological", or even the "physical level", as some kind of "filters", in the continuous activity of mutual auto-selection). And the idea of an immanentism (certainly, more methodological-operational and epistemological, than ontological) is, as it is well known, present in Hjelmslev, with the idea of the "principle of immanence"; and also, in direct continuation, with the development of textual semiotics, such as in Greimas.[8]

More generally, in Deleuze and Guattari's view the whole cosmos, therefore, appears pervaded by these "geo-semiotic" movements, certainly, no longer only verbal languages but all meaning production systems. While this generalization was already intrinsic to semiotic studies for some time, Deleuze and Guattari's theory seems to resemble a theory of "semiotic everywhere", a cosmo-theism made of layers (pre-significant semiotics, signifiers, post-signifiers, up to materials and technologies). Is there a risk in all of this of proposing a kind of theory of "every-thing"? We think not. The antidote lies precisely in Deleuze and Guattari's words: one must find specific planes of consistence, of immanence, and make them, cut them out and compose them. Finding areas of specific "application". This indication

[8] See, Greimas (1970; 1983; 1986; 1987). We refer, for a discussion of these points especially to the previously mentioned work of Bondì (2011, cit.) and Montanari (1997, 2012). With regard to the study of the principle of immanence in Hjelmslev, in Zinna (1986, 2008); see also Caputo (2010); a discussion of the points addressed by Zinna can be found in Montanari (cited above), particularly on the idea of a type of immanence that is methodological, meta-linguistic, and directed to a kind of "onto-logic": a study of the articulation of the layers. Greimas and Fontanille in *Sémiotique des Passions*, (1991; see also, Greimas and Courtès, 1979, 1986) claim: "the object of semiotics is phenomenological and paradoxically 'real' at the same time; from the point of view of the *ab quo* instance, semiotic existence of forms is of the 'manifested' order, the manifestation being the supposed and inaccessible "being"; from the point of view of the *ad quem* instance, semiotic forms are immanents, may be manifested in semiosis. Semiotic discourse is therefore a description of immanent structures and simulacra supposed reporting the conditions and preconditions of the manifestation of meaning, somehow, of 'being'" (my en. tr.): "(…) *l'objet de la sémiotique est phénoménal et paradoxalement 'réel' en même temps; du point de vue de l'instance ab quo, l'existence sémiotique des formes est de l'ordre du 'manifesté', la manifestation étant l''être' soupçonné et inaccessible; du point de vue de l'instance ad quem, les formes sémiotiques sont immanentes, susceptibles d'être manifestées lors de la sémiosis. Le discours sémiotique est dès lors la description des structures immanentes et des simulacres censés rendre compte des conditions et des pré-conditions de la manifestation du sens et, d'une certaine manière, de l''être'.*" (Greimas and Fontanille 1991, p. 11). For a critical discussion on this point, see also Coquet (1991; 2008); Marsciani and Zinna (1991); Marsciani and Pezzini (1997).

seems to come, at least in part, from general semiotics (but not generic) that made objects, technologies, as well as the verbal and visual, its own field of inquiry and discovery. We can think of "material semiotics" from scholars such as Law, that are based on Bruno Latour's research (see, Latour, cit.) and working on technologies and complex semiotic systems as well as, significantly, with high "index" of "de-territorialization", seeping, for instance, into disciplines such as economics and finance. Deleuze's theories have been partly "rediscovered" by semiotics in the last two decades, particularly through the work of Zilberberg (see 2002, 2006) and Fontanille (1995) (See also, Fabbri 1987; Guattari 1987). However, this discovery mainly focused on the question of the categories of "intensity" and tensive categories, which underlie oppositional categories, according to this hypothesis. But, perhaps, Deleuze and Guattari's work, coming from Spinoza (as well as Simondon, as we shall soon see), could have, could still be more strongly integrated when considering the statute of the semiotic discipline or some of its parts. In particular, in relation to "becoming" processes, the dynamics of the "infra-layer" relationships within complex models of production of signification.

## 10.7 Do We Really Need to Resist Simondon?

In conclusion we would like to return to Simondon, in order to address certain connections with our central theme of immanence and the plane of immanence (also in relation to semiotic systems), taking into account that Simondon is the other primary source of Deleuze's theories. The title of this section refers to Isabelle Stengers's essay, in which the scholar proposes to "*Résister à Simondon*" (while recognizing his importance to the development of Deleuze's thought) in order to assess the risk that Simondon's theory might be used as a kind of "*passe-partout*" concept, in particular regarding a concept like individuation. Stengers recognizes that

> The individuation process proposed by Simondon has explicitly tried to settle the question of the chicken and the egg dramatized by the human sciences: should we put at the principle of description or an individual with many well defined attributes or a functional environment from which properties that the individual thinks as its own will be deduced. For Simondon, the environment and the individual come "after" individuation, and if the description of one indicates the other, it is because a stable product of the process is proposed for the description in both cases. (my en. tr.):

> Le processus d'individuation proposé par Simondon a en effet pour visée explicite de trancher la question de la poule et de l'œuf que dramatisent tant de sciences humaines: faut-il mettre au principe de la description soit un individu doté d'attributs bien définis soit un milieu fonctionnel dont se déduiront les propriétés que l'individu pense siennes. Avec Simondon, et le milieu et l'individu viennent « après » l'individuation, et si la description de l'un renvoie alors à celle de l'autre, c'est parce que ce qui se propose à la description est dans les deux cas un produit stable du processus. (Stengers 2004, p. 55).

This is the point that Deleuze also addresses (as we saw earlier) regarding the break with the traditional concept of individuation, thanks to Simondon. What, however, Stengers stresses, quoting Combes (1999), as the critical point is the following:

> Simondon also created the concept of trans-individual, the more than individual, which means both the self-constitutive relationship of the subject and the meeting with a de-individuating event that, writes Muriel Combes (p. 66) 'breaks the functional inter-personal relationship and creates the need for proof.' Between auto-constitution (always already) and trans-individual dis-individuation as a condition of a new individuation, the experience (loneliness, anxiety) emerges which introduces the question of spirituality. It is there, it seems to me, that perhaps we must learn to resist Simondon. (my en. tr.):

> Simondon a également créé la notion de transindividuel, le plus qu'individuel, qui désigne à la fois ce qui est relation auto-constituante du sujet et ce dont la rencontre est l'événement désindividuant qui, écrit Muriel Combes (p. 66), 'brise le rapport interindividuel fonctionnel et engendre la nécessité d'une épreuve'. Entre auto-constitution (toujours déjà) et désindividuation transindividuelle comme condition d'une nouvelle individuation, se dessine ainsi la place d'une expérience (solitude, angoisse) qui introduit à la question de la spiritualité. C'est là, me semble-t-il, qu'il faut peut-être apprendre à résister à Simondon. (Stengers, ib., p. 57).

Therefore, the risk, as Stengers points out, is found even in the use of concepts such as trans-individual and collective individuation (used, moreover, by scholars such as Balibar—see, Balibar, Morfino (2014)—as well as Deleuze and Guattari) in the direction of a sort of "spiritualism" or as a pre-individual "neo-humanism", due precisely to a line of thought, that seems to find phenomena, such as "transduction" (phenomenon for Simondon, "physical, biological, mental" at the same time) everywhere. The antidote, according to Stengers, is Deleuze and Guattari: "find the ways" to play (assemble, therefore) Deleuze and Guattari's thought "against" Simondon. In this sense resisting him. And, we might add, based on what is written above, even contemplating Deleuze and Guattari's immanence means "finding cases', resisting the temptation of this "passe-partout" and "explanation of everything".

In any case, if we agree with Stengers's concerns, one might say that this has somehow "already happened". In the sense that Deleuze and Guattari's theories seemed to have worked on Simondon's concepts in this direction. We will quickly address certain points of this relationship, and the concepts developed by Simondon that were later taken up and articulated by Deleuze and Guattari. First of all, we have already mentioned, the new idea of individuation, which, for Deleuze (1964) assumes an ethical dimension (we could say, precisely in Spinoza's sense). The pre-individual remains associated with individuals, Deleuze says, but precisely as a *place of production*, as process, place of the possibilities ahead of the individuals themselves and their individuation and transformations. There is here, as Deleuze points, Simondon's other major concept, metastable systems: pre-individuality as a source of metastable states for the future. That is the sense in which, says Deleuze, Simondon's concept of information becomes ethical: the being, the self, and subjectivity are always metastable places, even once they are "identified". Therefore, able to "become another". The idea of potential, defined as the possibility for variation in systems and in the transformation of system into other thing. Simondon's other fundamental concepts follow from this, such as the invention of a new matter/form relationship through the rejection of the traditional Aristotelian hylomorphism. The replacement of this relationship with other concepts, such as those of the mould, material, forces and milieus from which it is possible to "make new forms" (the well-known example of the artisan-potter, for which objects are never independent, Simondon stresses, but they are always considered as networks of collective relations, almost terminals of these same relationships, represents an anticipation of more recent socio-semiotic analysis of techniques, such as in Latour (see, i.e., Latour 1991, 2014)). Here we find Deleuze and Guattari's proposal, regarding what was emphasized above in relation to Hjelmslev's semiotic system model. We must replace, as Deleuze and Guattari say *in Thousands of Plateaus*, the categories (in relation to semiotic layers) of "matter" and "form" with simondonian "materials" and "forces". Assess the dynamics that lead to transformations between layers. Think about the very idea of *transduction* (see Simondon 1958a, p. 34; see also, Simondon 1958b) as *one* of the possible processes that intervene between the layers, in the production of material and immaterial meaning.

Transduction, as long as it is thought of in a specific way, becomes the dissemination model, of a modification of different regions of a system, as a "structuring" mechanism. Simondon attempts to link topological and spatial models with noetic models (of form construction), that seem to have been taken up and developed, once again, by Deleuze and Guattari. Transduction, Simondon stresses, does not look "elsewhere" (ib.) for a principle to solve a problem in a given domain, but obtains its "resolutive structure" from the domain itself. It seems to us that, even if Simondon can be interpreted as "at risk of" a certain 'spiritualism'", and as much as he himself is critical of Spinoza's theories, in this case he offers us an excellent definition of what is the expressive immanence. According to this conception it is important not looking for an external form, rather look in the cracks and fault lines of the systems themselves, the formations and forces of meaning creation. Forms and forces of the production of meaning that act like processes, assembled streams, but, at the same time, as multiple events. From which, once again, to extract, and find heterogeneous orders that enter into a relationships with each other. The pot and the potter.

As Simondon says (ibid., p. 29): "The psycho-social worlds of the transindividual is neither the crude nor interpersonal social; it requires effective operation of individuation starting from a preindividual reality, associated with individuals and capable of forming a new problem having its own metastability; it expresses a quantum condition, corresponding to a plurality of orders of magnitude. The living is presented as a *problematic thing*, superior and inferior at the same time. Saying that living is problematic, is to considered becoming as a dimension of life: the living depends on becoming, which acts as a mediation." (my en. tr.):

> Le monde psycho-social du transindividuel n'est ni le social brut ni l'interindividuel; il suppose une véritable opération d'individuation à partir d'une réalité préindividuelle,

associée aux individus et capable de constituer une nouvelle problématique ayant sa propre métastabilité; il exprime une condition quantique, corrélative d'une pluralité d'ordres de grandeur. Le vivant est présenté comme être problématique, supérieur et inférieur à la fois à l'unité. Dire que le vivant est problématique, c'est considérer le devenir comme une dimension du vivant: le vivant est selon le devenir, qui opère une médiation.

## References


Alquié, F.: Le rationalisme de Spinoza. Puf, Paris (1981)

Althusser L.: tr. it., L'unica Tradizione Materialista: Spinoza. CUEM, Milano (1998)

Assman, J.: Moses the Egyptian: The Memory of Egypt in Western Monotheism. Harvard University Press, Cambridge (1997)

Balibar, E., Morfino, P. (eds.): Il Transindividuale. Mimesis, Milano (2014)

Banfi, A.: Spinoza e il suo Tempo. Vallecchi, Firenze (1969)

Cristofolini, P.: Spinoza e la gioia", In: AA.VV., L'eresia della libertà, Omaggio a Paolo Cristofolini. ETS, Pisa (2008)

Balibar, E.: Spinoza et la Politique. PUF, Paris (1985)

Bondì, A.: La Parola e i suoi Strati. La Semiotica Dinamica di Louis Hjelmslev. Bonanno, Catania (2011)

Caputo, C.: Hjelmslev e la Semiotica. Carocci, Roma (2010)

Ciccarelli, R.: Immanenza. Il Mulino, Bologna (2008)

Combes, M.: Simondon. Individu et Collectivité. PUF, Paris (1999)

Coquet, J-Cl.: Realité et Principe D'immanence", In: Coquet, J-Cl., Petitot, J., (éds.), Langage, sens et realité, Langages, vol. 103, Larousse, Paris (1991)

Coquet, J-Cl: Le istanze Enuncianti. Bruno Mondadori, Milano (2008)

Cross, R.: Medieval theories of haecceity. In: Edward, N.Z. (ed.) The Stanford Encyclopedia of Philosophy (Fall 2010 Edition), : http://plato.stanford.edu/archives/fall2010/entries/medieval-haecceity/ (visited online: march 2014) (2010)

Damasio, A.: Looking for Spinoza: Joy, Sorrow and the Feeling Brain. Harcourt, Orlando (2003)

De Gaudemar, M.: Lecture et appareil critique. In: Leibniz, G.W. (ed.) Réfutation inédite de Spinoza, Actes Sud Arles (1999)

Deleuze, G.: Gilbert Simondon, L'individu et sa génèse physicho-biologique. Introduction. In: L'Ile déserte et autres textes, Minuit, Paris 2002 (1964)

Deleuze, G.: Différence et Répétition. PUF, Paris (1968a)

Deleuze, G.: Spinoza et le Problème de l'Expression. Minuit, Paris (1968b)

Deleuze, G.: Spinoza. Philosophie Pratique. Minuit, Paris (1981)

Deleuze, G.: Le Pli. Minuit, Paris (1988)

Deleuze, G.: Interview. Magazine Litteraire vol. 257, Paris (1988)

Deleuze, G.: Critique et Clinique. Minuit, Paris (1993)

Deleuze, G., Guattari, F.: Mille Plateaux. Minuit, Paris (en. tr.: A Thousand of Plateaus, Minneapolis: University of Minnesota Press, 1987) (1980)

Deleuze, G., Guattari, F.: Qu'est-ce-que la philosophie?. Minuit, Paris (en. tr.: *What is Philosophy*, New York: Columbia University Press, 1994) (1991)

Deleuze, G.: L'immanence: une vie. Minuit, Paris (1995)

Deleuze, G.: Cosa può un corpo? Lezioni su Spinoza. Ombre corte, Verona (2007)

Dosse, F.: Gilles Deleuze et Félix Guattari. Biographie croisée. La Découverte, Paris (2007)

Dawkins, R.: It took Spinoza and structuralism to teach Deleuze that meaning is not necessarily attributed to the cinematic sign. In: Semiotica, vol. 157, de Gruyter, Berlin (2005)

Duffy, S.: The Logic of Expression; Quality, Quantity, and Intensity in Spinoza, Hegel and Deleuze. Ashgate, Aldershot (2006)

Fabbri, P.: A Passion veduta. Il vaglio semiotico. In: Fabbri, P., Pezzini, I. (eds.) Affettività e sistemi semiotici. Le passioni nel discorso, Versus, p. 47/48, Bompiani, Milano (1987)
Fabbri, P.: Come Deleuze ci fa segno. Da Hjelmslev a Peirce. In Vaccaro, S. (ed.) Il Secolo Deleuziano, Mimesis, Milano (1997)
Feldman, S.: Introduction, Spinoza, B., Ethics. The Treatise on The Emendation of the Intellect and Selected Letters, Hackett Publ. Company, Indianapolis (1992)
Fontanille, J. (ed.): Le dévenir Actes du Colloque. Pulim, Limoges (1995)
Greimas, A.J.: Du Sens. Seuil, Paris (1970)
Greimas, A.J.: Du Sens II. Seuil, Paris (1983)
Greimas, A.J.: De la nostalgie. Etude de sémantique lexicale. In: Actes Sémiotiques, Bulletin, GRSL, Paris (1986)
Greimas, A.J.: De l'imperfection. P. Fanlac, Paris (1987)
Greimas, A.J., Courtés, J.: Sémiotique. Dictionnaire raisonné de la theorie du langage, Tomes I et II Hachette, Paris (1979, 1986)
Greimas, A.J., Fontanille, J.: Sémiotique des Passions. Des états de choses aux états d'âme. Seuil, Paris (1991)
Guérin, M.: L'affectivité de la pensée. Actes Sud, Arles (1992)
Guattari, F.: Ritournelles et affects existentiels, In: Fabbri, P., Pezzini, I. (eds.) (1987)
Hjelmslev, L.: Prolégomènes à une théorie du langage. Minuit, Paris (éd. or.: 1943) (1968)
Hjelmslev, L.: Essais linguistiques. Minuit, Paris (1971)
Israel, J.I.: Radical Enlightenment: Philosophy and the Making of Modernity 1650–1750. Oxford University Press, Oxford (2001)
Latour, B.: Nous n'avons jamais été modernes. Essai d'anthropologie symétrique. La Découverte, Paris (1991)
Latour, B.: Reflections on Etienne Souriau's Les différents modes d'existence. http://www.bruno-latour.fr/sites/default/files/downloads/98-SOURIAU-GRAHAM-GB.pdf (2014)
Leibniz, G.W.: Réfutation inédite de Spinoza, a cura e con nota introduttiva di M. De Gaudemar (1999)
Luhmann, N.: Sistemi sociali. Il Mulino, Bologna (éd. or.: 1984, *Soziale Systeme*, Suhrkamp Verlag, Frankfurt am Main) (1990)
Matheron, A.: Individu et communaute chez Spinoza. Minuit, Paris (1969)
Marsciani, F., Zinna, A.: Elementi di semiotica generativa. Esculapio, Bologna (1991)
Marsciani, F.: Ricerche semiotiche I. Il tema trascendentale. Esculapio, Bologna (2012)
Marsciani, F., Pezzini, I.: Introduzione. In: Greimas, F. (ed.) (1991, trad. it) (1997)
Montanari, F.: Between trees, webs and mirrors. Dimensions of Immanence and a critical poststructuralist proposal", In: E/C: www.ec-aiss.it (2012)
Montanari, F.: Esprimere l'immanenza. In: Vaccaro, S. (ed.) Il secolo deleuziano. Milano, Mimesis (1997)
Nadler, S.M.: Spinoza: A life. Cambridge University Press, Cambridge (1999)
Nadler, S.M.: Spinoza's Heresy. Oxford University Press, Oxford (2001)
Negri, A. (ed.): Spinoza: l'anomalia selvaggia, Spinoza sovversivo, democrazia ed eternità in Spinoza. DeriveApprodi, Roma (1998)
Ouellet, P.:Signification et sensation. Nouveaux Actes Sémiotiques, vol. 20, PULIM, Limoges (1992)
Patton, P.: Deleuze: A Critical Reader. Blackwell, Oxford (1996)
Petitot-Cocorda, J.: Morphognèse du sens. PUF, Paris (1985)
Petitot-Cocorda, J.: (in collaboration with Doursat, P.) Cognitive Morphodynamics: Dynamical Morphological Models of Constituency in Perception and Syntax. Peter Lang, Berne (2011)
Portela, J.Cr.: Conversations avec Jacques Fontanille. Alfa, São Paulo, vol. 50(1), pp. 159–186 (2006)
Sauvagnargues, A.: Gilbert Simondon. In: Leclercq, S. (ed.) Aux sources de la pensée de Gilles Deleuze. Sils Maria, Mons, Paris (2005)
Simondon, G.: (1958a) L'individuation à la lumière des notions de forme et d'information. Millon, Grenoble (2005)

Simondon, G.: (1958b, 1989) Des modes d'existence des objects techniques. Aubier, Paris (2001)
Spinoza, B., 1972, 1991 (2nd ed.), Etica. Trattato teologico-politico, [1675–1677], Torino, TEA-UTET (versione italiana a cura di Remo Cantoni e Franco Fergnani, condotta sulla edizione critica e traduzione delle opere complete di Spinoza curata da Carl Gebhardt, Heidelberg, 1923–1926, a partire dall'edizione del 1677). For references in english, see: Spinoza, B., *Ethics. Treatise on the Emendation of the Intellect and Selected Letters*, translated by Samuel Shirley, edited and introduced by Seymour Feldman, Hackett Publishing Company, Indianapolis (1992)
Stengers, I.: Résister à Simondon?. In: Multitudes, vol. 18, pp. 55–62 (2004)
Wolfson, H.A.: The Philosophy of Spinoza. Harvard University Press, Harvard (1934)
Yovel, Y.: Spinoza and Other Heretics. Princeton University Press, Princeton (1989)
Zilberberg, C.l.: Précis de Grammaire Tensive", *Tangence*, Rimouski, vol. 70, p.111–143 (2002)
Zilberberg, Cl: Eléments de Grammaire Tensive. PULIM, Limoges (2006)
Zinna, A. (ed.): Louis Hjelmslev. Linguistica e semiotica strutturale", Versus, vol. 43, gennaio-aprile (1986)
Zinna, A.: Il primato dell'immanenza nella semiotica strutturale In: E/C: www.e/c-aiss.it (2008)

MIX
Papier aus verantwortungsvollen Quellen
Paper from responsible sources
FSC® C105338

If you have any concerns about our products,
you can contact us on
**ProductSafety@springernature.com**

In case Publisher is established outside the EU,
the EU authorized representative is:
**Springer Nature Customer Service Center GmbH**
**Europaplatz 3, 69115 Heidelberg, Germany**

Printed by Libri Plureos GmbH
in Hamburg, Germany